Geotours Workbook

A Guide for
Exploring Geology using Google Earth

Second Edition

M. SCOTT WILKERSON
Department of Geosciences
DePauw University

M. BETH WILKERSON
Library & Information Services
DePauw University

STEPHEN MARSHAK
Department of Geology
University of Illinois

W. W. Norton & Company
New York • London

W. W. Norton & Company has been independent since its founding in 1923, when William Warder Norton and Mary D. Herter Norton first published lectures delivered at the People's Institute, the adult education division of New York City's Cooper Union. The Nortons soon expanded their program beyond the Institute, publishing books by celebrated academics from America and abroad. By mid-century, the two major pillars of Norton's publishing program—trade books and college texts—were firmly established. In the 1950s, the Norton family transferred control of the company to its employees, and today—with a staff of four hundred and a comparable number of trade, college, and professional titles published each year—W. W. Norton & Company stands as the largest and oldest publishing house owned wholly by its employees.

ISBN: 978-1-324-00096-9 (pbk.)

W. W. Norton & Company, Inc., 500 Fifth Avenue, New York, NY 10110
wwnorton.com

W. W. Norton & Company, Ltd., Carlisle Street, London W1D 3BS

9 8 7 6 5 4 3 2 1

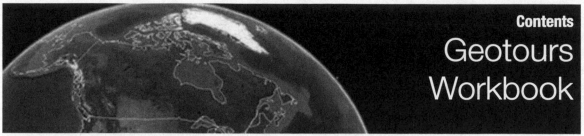

Contents
Geotours Workbook

Section 2: Exploring Geology Using Geotours

Section 3: Developing Your Own Geotours

Preface

If you're reading this, you probably (a) already have spent numerous hours exploring the seductive virtual world of Google Earth and yearn to learn more about the software and the planet on which we live or (b) haven't yet used Google Earth (perhaps you're taking a course that requires you to use this workbook) and are on the brink of realizing how empty your life has been up to this point! In either case, sit back—your world is about to be "rocked."

Google Earth is a free "virtual globe" that simulates the planet on which we live by draping high-resolution global satellite imagery onto a sphere with 3-D terrain—*over the entire planet*. With a click of a mouse, users can take interactive virtual field trips anywhere in the world to explore the Earth's surface (even the ocean floor). Using the interactive 3-D interface, users literally can zoom and fly around to visualize any region on the planet from any direction, distance, and angle. As if that isn't enough, Google Earth also has the tools to be a "geographic web browser" to associate text, pictures, and movies with specific spatial locations on the virtual globe. There's more—lots more, but you don't want to learn everything about Google Earth in the preface, do you?

This workbook started life back in 2008 (yes, books do take on a life of their own...). What originated as a series of placemarks (aka Geotours) to automatically whisk users to various locations all over the planet to highlight cool geologic features quickly morphed into a textbook supplement with worksheets to help students better understand what they were seeing. Predictably, with time (and new textbook editions), the Geotours and worksheets multiplied. Inevitably, users also wanted to know how to make their own Geotours and/or how to teach their students to create Geotour projects (hence, the workbook that you have before you).

You may be asking—*How should I use this book?* That depends on what you want to do...

All users

All users should read **Section 1**, which provides a basic introduction to (1) downloading and installing Google Earth and the files associated with this workbook and (2) using Google Earth.

Users who only want to learn about the Earth

Users who only want to learn about the Earth (e.g., you are taking an earth science or geology class) should focus on the Geotour Worksheets in **Section 2** and the associated Geotour KMZ files. *The worksheets are meant to supplement a geoscience textbook, so it is important to refer to such a textbook and/or similar geoscience reference materials (online or print) for background information in order to complete the worksheets.*

For classroom use, worksheets typically are assigned as student homework or in-class assignments (see Table 1 for academic users who have adopted *Earth: Portrait of a Planet or Essentials of Geology* by Stephen Marshak). Alternatively, instructors may want to use materials in the Geotour KMZ files during lectures and/or to direct interested students to these materials for self-exploration.

Table 1. Using *Geotours Workbook* with *Earth: Portrait of a Planet* or *Essentials of Geology* by Stephen Marshak (W. W. Norton).		
Section 2 Geotour worksheet	*Earth: Portrait of a Planet* chapter	*Essentials of Geology* chapter
A	1	1
B	2–4	2
C	5	3
D	6	4
E	9	5
F	7	6
G	8	7
H	10	8
I	11	9
J	12	10
K	13	11
L	14–15	12
M	16	13
N	17	14
O	18	15
P	19	16
Q	21	17
R	22	18
S	23	19

Users who only want to learn how to create Google Earth materials

Users who only want to learn how to create materials for Google Earth can go directly to **Section 3**. This section provides step-by-step instructions for creating Google Earth materials (aka "Geotours") without droning on with extensive text explanations. As such, our approach is to divide each module into "projects" that generally contain less than 10 steps and have a single focused objective in mind. That is, in the Placemarks module, we create a placemark in one project, edit the placemark in a second project, format the text in a third project, add an image in a fourth project, and so on...you get the idea.

We think this approach should work as well in a classroom with a wide range of student abilities as it does as a workbook for individuals who want to work through the material on their own. Examples are heavily geared toward geology, earth science, environmental science, etc., but the techniques are applicable to a wide array of disciplines.

Users who want to learn about the Earth and then create their own geoscience-based Geotour projects or other Google Earth materials

We like you! We would recommend that you at least work through some of the worksheets in **Section 2** before starting **Section 3** (the more of **Section 2**, the better). That way, you not only obtain some exposure to geoscience topics in the context of Google Earth, but you also gain an appreciation for the range of Google Earth's capabilities.

We've used this approach in the classroom, and it works quite well. Typically, we assign topics in **Section 2** throughout the semester and then ask students to complete a Geotour project as a final semester project using **Section 3**. We typically work through **Section 3** as a group during a couple of lab periods (although most students are quite capable of working through the projects on their own).

New in the Second Edition
Thanks so much to those of you who provided feedback and suggestions for new and exciting Geotours and ways to use them! We've been very busy, and we hope that our improvements in this second edition make the *Geotours Workbook* even more effective for how faculty teach and how students learn geoscience concepts. Specifically, we have...

- **Updated the book to be compatible with the Google Earth Pro 7.1.7 interface and functionality** (the first edition of the book was compatible with Google Earth 6.1). The move to Google Earth Pro provides several advanced tools (e.g., the ability to measure areas) that are incredibly useful for geoscientists. And, best of all, Google Earth Pro is now *free!*

- **Streamlined the Google Earth KMZ file.** We reorganized the Google Earth KMZ file structure to better reflect how faculty teach and how students work. That is, the Geotours library of examples for each topic is now readily accessible from the same folder as the worksheet problem placemarks, allowing students to easily explore Geotours of related features as they work on the worksheet problems. Additionally, faculty can access the Geotours for examples while they are lecturing without having to drill down through layers of folders.

- **Revised and reformatted the Geotour worksheet questions.** Some worksheets have been totally overhauled, whereas other worksheets have seen mostly minor tweaks.

  For example...
  - **Geotour Worksheet A: Earth & Sky** now helps students explore the solar system by scaling the solar system (e.g., size of the Sun and the planets, distances between planets, etc.) to the distance between Los Angeles and New York City and by exploring features on Google Moon and Google Mars.

  - **Geotour Worksheet C: Minerals** now includes a totally revised series of relevant and applied questions pertaining to minerals...even showing students the environmental effects on the areas surrounding where minerals for our mobile phones are mined.

  - **Geotour Worksheet H: Earthquakes** now allows students to use Google Earth Pro's capabilities to not only locate the epicenter of an earthquake from seismograms, but also to calculate its Richter magnitude.

  All worksheets have been reformatted to place important instructions for the students on the first page before the questions and to include a header for each question that provides not only the Geotours geographic location, but also the geoscience topic that the question addresses. Based on our experience teaching with the *Geotours Workbook*, these small formatting changes can reap large benefits in terms of student engagement and comprehension.

- **Added MANY new Geotour locations to the Geotour Worksheet libraries.** We have greatly expanded both the number and geographic diversity of Geotour sites housed in each Geotour Worksheet library. Specifically, we spent a great deal of time not only conducting research for classic and currently relevant examples, but also using Google Earth to fly over every single continent looking for spectacular landform examples (we know...tough job, right?). For example, the **Volcanoes Geotours** library now not only has folders for each type of volcano, but also has subfolders divided by continent that include Geotours to famous volcanoes and lesser known volcanoes that exhibit remarkably ideal or unique characteristics.

- And much more!

Our goal is that this workbook helps you to explore important interdisciplinary geoscience topics in their proper spatial context, to better recognize geologic landforms and environmental issues affecting our planet, and to attain a basic knowledge of how these features and issues have formed and evolved. In addition, we hope that the workbook helps you develop the technical skills you need to prepare your own Google Earth materials for your own projects. Enjoy!

Acknowledgments

We're very grateful for the assistance and support of numerous individuals and organizations during the development of both editions of this book. First, this book would not have been possible without the exceptional support of DePauw University. Specifically, we would like to thank Carol Smith, Neal Abraham, Fred Soster, Jim Mills, Tim Cope, and Jeane Pope for their help/feedback, support, advice, and encouragement. Second, Jack Repcheck, Eric Svendsen, Jake Schindel, Rob Bellinger, Matthew Freeman, and many other W. W. Norton employees provided important logistical and editorial help, advice, and guidance, without which this book would never have been published. In addition, various individuals and organizations listed in the Credits and in the KMZ file were very gracious to share photos, figures, diagrams, KML data, etc., with us. The second edition of this book has greatly benefitted from reviews by Cara Burberry, Cinzia Cervato, Melinda Hutson, and Alexander Stewart and from numerous suggestions for Geotour sites and improvements from students and adopters. Finally, Zach and Ben Wilkerson were excellent companions while Scott and Beth worked on this project...more than willing to work through exercises, never complaining about the long hours at the office, and always providing love and moral support.

About the Authors

Scott Wilkerson *(mswilke@depauw.edu)* is a Professor of Geosciences at DePauw University in Greencastle, IN. He holds a B.S. from Murray State University and a Ph.D. from the University of Illinois, Urbana-Champaign. Prior to his appointment at DePauw, Scott was a senior research geologist with Exxon Production Research in Houston, TX. Scott's research focus is understanding the 2-D and 3-D geometric and kinematic development of fault-related folds (using field mapping, computer/analog modeling, and cross-section balancing). He also is interested in visualizing topographic landforms and structural features using Google Earth and other software programs. Scott has been named a Faculty Fellow and a University Professor at DePauw University and currently holds the Ernest R. Smith Professor of Geosciences endowed chair.

Beth Wilkerson *(bwilkerson@depauw.edu)* is the GIS Specialist at DePauw University. She holds a B.S. from Murray State University and an M.S. from the University of Illinois, Urbana-Champaign in applied mathematics. Prior to becoming the GIS Specialist at DePauw, Beth was a senior software engineer at Schlumberger-Geoquest, where she specialized in programming 2-D and 3-D interpretation and visualization systems for the geophysical services industry. Beth currently supervises DePauw's GIS Center, where she works with faculty, staff, students, and administrators on GIS projects involving teaching, research, and university initiatives. Beth and Scott are married and have two children, Zach and Ben.

Stephen Marshak *(smarshak@illinois.edu)* is a Professor of Geology and Director of the School of Earth, Society, & Environment at the University of Illinois, Urbana-Champaign. He holds an A.B. from Cornell University, an M.S. from the University of Arizona, and a Ph.D. from Columbia University. Steve's research interests center on a wide range of topics involving structural geology and tectonics. Steve has been recognized for his outstanding teaching by receiving several university teaching awards and the 2012 Neil Miner Award from the National Association of Geoscience Teachers for "exceptional contributions to the stimulation of interest in the Earth sciences." In addition, Steve is the author of *Earth: Portrait of a Planet* and *Essentials of Geology* and is the co-author of *Earth Structure: An Introduction to Structural Geology and Tectonics*, *Basic Methods of Structural Geology*, and *Laboratory Manual for Introductory Geology*.

Getting Started

Google Earth

Imagery Date: 12/14/2015   lat 31.890450° lon -97.827285° elev 307 m   eye alt 8076.86 km

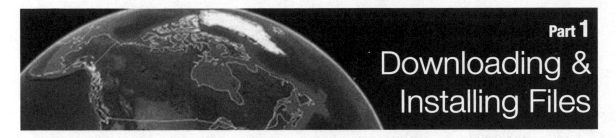

Part 1

Downloading & Installing Files

In this part, you will learn how to download the Google Earth application and related *Geotour Workbook* materials.

Topic 1.1	downloading the Google Earth application

1. Using your web browser, go to the following website: http://earth.google.com/.

2. Click on the Google Earth Pro box. On the webpage that appears, you may download Google Earth (free) or Google Earth Pro (free since January 2015). Click on the "**Download Google Earth Pro**" button to review the license agreement and to begin downloading (the site automatically downloads the appropriate version for your computer). For a comparison of the two versions, go to https://support.google.com/earth/answer/189188?hl=en.

> **What version of Google Earth was used to create this workbook?**
>
> This workbook was created using **Google Earth Pro 7.1.7**. Please note that the Google Earth application is constantly changing, and so interface elements, imagery quality, and layer content included with Google Earth may look somewhat different from what is shown in this workbook.

3. Follow the on-screen instructions to install Google Earth Pro on your particular system. The installer usually defaults to installing the application in the folder that contains all of your applications. If you are asked for a username and license code when you first run the application, please use your **email address** as your username and **GEPFREE** for the license code.

Topic 1.2	downloading the *Geotours Workbook* KMZ file

1. Using your web browser, download the following file to your desktop:
   http://media.wwnorton.com/college/geo/geotours2/Geotours2e.kmz

2. Double-click the **Geotours2e.kmz** file on your desktop, and Google Earth typically will launch automatically (or select **File > Open > Geotours2e.kmz** from within Google Earth).

3. In the left-hand sidebar you will see a **Places** panel, and in the **Temporary Places** folder you will see a **Geotours2e** network link. Drag the **Geotours2e** network link into the **My Places** folder.

4. Click on the box next to the **Geotours2e** network link to download the latest **GeotoursWorkbook2e** folder. Click on the triangle to open the **Geotours2e** network link to see the **GeotoursWorkbook2e** folder. *Note: Each time you quit & restart Google Earth, you will repeat step 4 in order to automatically download the latest **GeotoursWorkbook2e** folder.*

5. Click on the triangle next to the **GeotoursWorkbook2e** folder to expand the folder and reveal its contents:

> **Welcome**
> **Section 1**. *Getting Started*
> **Section 2**. *Exploring Geology Using Geotours*
> **Section 3**. *Developing Your Own Google Earth Geotours*

6. The **Welcome** folder contains selected **Sample Geotours** that showcase some of Google Earth's capabilities (grouped into folders based on the Google Earth feature that they highlight). Begin your exploration by first clicking on the folder name to see a brief description of that Google Earth feature. Next, check and double-click on each example Geotour within the folder to take a "tour" of some fantastic geo-location. It is our hope that this activity will not only help you realize the value and power of Google Earth as a means of better understanding the planet on which we live, but that it also helps whet your appetite for what is to come!

7. The **Section 1** folder provides instructions on how to (1) download and install Google Earth and the files associated with this workbook and (2) use Google Earth.

8. The **Section 2** folder contains Geotour Worksheet subfolders arranged by topics generally covered in most introductory geoscience textbooks. Each topical folder includes:

    a. **Google Earth materials (e.g., placemarks, photos, overlays, etc.) keyed to each Geotour Worksheet question** in Section 2 of the *Geotours Workbook*.

    b. **a green "[topic] Geotours Library" folder**. This folder provides a comprehensive, one-stop library of all available Geotours that are germane to that particular geoscience topic. Rather than printing a static list/index of Geotours in the *Geotours Workbook*, we encourage users to interactively explore the digital libraries for each topic in the *Geotours Workbook* KMZ file. Such an approach not only is more effective than thumbing through hard-copy pages in the workbook (especially if coupled with the *Find in My Places* search feature in the *Places* panel), but it also allows us to periodically update the Geotour libraries with more content.

9. The **Section 3** folder provides files illustrating how to develop Google Earth Geotour content.

10. Enjoy!

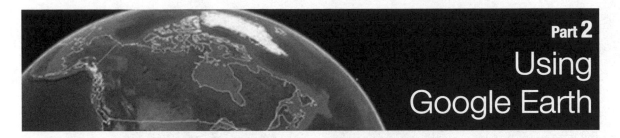

In this part, you will learn how to use the Google Earth interface. In addition, you will learn about many of Google Earth's powerful features and the content that is provided free with the application. *Please note that SCROLL = mouse scroll wheel, LMB = left mouse button, and RMB = right mouse button. Unless specified, "click" refers to LMB. If major parts of the Google Earth interface (e.g., toolbar, panels sidebar, status bar, navigation controls, etc.) are hidden, these can be readily turned on using the View menu.*

Topic 2.1	**the toolbar**

When you start the Google Earth application, the program will initialize and an image of the Earth with a background of stars will appear on your screen. A **toolbar** across the top of the viewer window (Figure 2.1) provides **icons** that are quick links to performing various tasks.

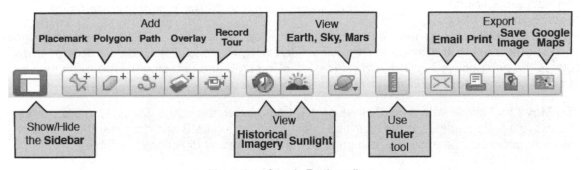

Figure 2.1: Google Earth toolbar.

Is the imagery gathered in real time?

No, Google Earth is powerful, but not THAT powerful! The imagery date is usually shown along the information bar at the bottom of the viewer window (described in the following pages). Google Earth tries to keep its current imagery updated within the past few years (while providing access to available historical imagery as well).

Topic 2.2 | the navigation controls

In Google Earth, navigation tools appear in the upper right-hand corner of the screen (Figure 2.2; **View > Show Navigation > Automatically**).

The **Look tool** (circle with an eye in the center) allows you to rotate the view. You can do this either by clicking LMB and dragging the mouse on or within the outer ring, or by rotating the view clockwise or counterclockwise by clicking on the left or right arrows, respectively. Clicking (or dragging) the N

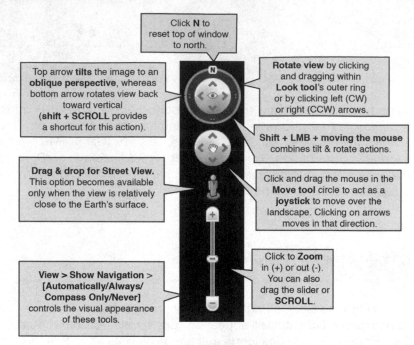

Click **N** to reset top of window to north.

Top arrow **tilts** the image to an **oblique perspective**, whereas bottom arrow rotates view back toward vertical (**shift + SCROLL** provides a shortcut for this action).

Rotate view by clicking and dragging within **Look tool**'s outer ring or by clicking left (CW) or right (CCW) arrows.

Shift + LMB + moving the mouse combines tilt & rotate actions.

Drag & drop for Street View. This option becomes available only when the view is relatively close to the Earth's surface.

Click and drag the mouse in the **Move tool** circle to act as a **joystick** to move over the landscape. Clicking on arrows moves in that direction.

View > Show Navigation > [Automatically/Always/ Compass Only/Never] controls the visual appearance of these tools.

Click to **Zoom** in (+) or out (-). You can also drag the slider or **SCROLL**.

Figure 2.2: Google Earth navigation controls.
For additional information, see the Google Earth User Guide
(http://earth.google.com/support/bin/static.py?page=guide_toc.cs).

button with LMB reorients the view with north at the top of the screen. Clicking on the top arrow of the **Look tool** tilts the image to show an oblique view, whereas clicking on the bottom arrow rotates the perspective back toward vertical. Clicking LMB and dragging the mouse in the inner circle of the **Look tool** performs **both** actions simultaneously.

The **Move tool** (circle with a hand in the center) pans across the image. By clicking the arrows, you move a finite distance in the specified direction. Clicking LMB and dragging the mouse in the inner circle acts as a joystick and translates the view in any direction (see Table 2.1 for mouse shortcuts).

Table 2.1. mouse shortcuts			
	LMB (left mouse button)	**Scroll Wheel**	**RMB** (right mouse button)
No Key	click/hold to pan & click/drag/ release to fly over area	zoom in/out	rotate view & zoom in/out simultaneously while moving mouse
	double-click to zoom in		double-click to zoom out
Shift Key	rotate view & tilt view to/from perspective simultaneously while moving mouse	tilt view to/from perspective	rotate view & zoom in/out simultaneously while moving mouse

As you zoom toward the ground surface, the **Street View tool** appears in the navigation controls (Figure 2.2, person icon). Dragging the person icon to a location in the Google Earth window will highlight nearby roads in blue if Street View photography is available. Placing the person icon along one of these roads will zoom and tilt the view to provide a ground-level perspective. The screen will enter Street View mode with Street View photography on the screen and information/controls in the upper right-hand corner (e.g., the street address, a person button highlighted in blue, and the **Exit Street View button**; Figure 2.3). To toggle from Street View to ground-level view, click the building button next to the person button. The building button will highlight in blue, the address will disappear, and the Exit Street View button changes to **Exit ground-level view**. Note that the ground-level mode will appear if the person icon is placed at a location not associated with a road (or if the road has no Street View photography).

The vertical **Zoom slider** zooms the view up or down. You can either move the slider or click on the end of the bar (see Table 2.1 for mouse shortcuts). Clicking on the + end takes you to a lower elevation, whereas clicking on the – end takes you to a higher elevation. The slider offers very fine control of zooming. Select **View > Scale Legend** from the menu to display a bar scale in the lower left portion of the screen. The default setting in Google Earth is to automatically swoop (tilt) into a perspective view as you zoom closer. This auto-tilt behavior may be toggled on and off in the **Preferences (Mac)/Options (PC)** dialogs.

Figure 2.3: Google Earth Street View/ground-level view controls. Click on an object in Street View to advance to that object. RMB <u>or</u> Control-LMB followed by moving the cursor down-up will zoom in and out of the Street View imagery, respectively. SCROLL moves the view along the street.

On the Earth image, you will see a hand-shaped cursor. By clicking and dragging the cursor across the screen, you can move the image. If you quickly drag the hand cursor while holding down the mouse and then let go, the movement will continue (first make sure to uncheck **Gradually slow the Earth when rotating or zooming** in **Preferences > Navigation**). Clicking anywhere in the viewer or hitting the spacebar will stop this continual motion of the Earth.

Topic 2.3	the status bar

On the bottom of a Google Earth window, you will see several status items (Figure 2.4). On the left side of the screen, you will find a **scale bar** for the view (toggle on/off using **View > Scale Legend**), a **Tour Guide** with nearby sites of interest (toggle on/off using **View > Tour Guide**), and if zoomed in enough, a **Historical Imagery link** with the date of the oldest imagery available in Google Earth.

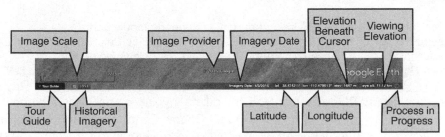

Figure 2.4: Google Earth status bar.

On the right side of the screen (L to R), you will see the **date of the imagery** in the window. Next, you will see the **location coordinates** and **ground surface elevation** of the point just beneath the hand-shaped cursor on the screen. The location coordinate units default to **latitude** (north [+] or south [-] of the equator) and **longitude** (east [+] or west [-] of the prime meridian). The coordinate system and format can be changed in the **Preferences (Mac)/Options (PC)** dialogs. Two versatile choices are decimal degrees (latitude/longitude) or Universal Transverse Mercator (a metric-based coordinate system referenced to latitude/longitude). On the far right side, you will see the height of the **viewing elevation** ("Eye alt") and a **blue circle** that graphically shows if your computer is busy processing (e.g., streaming image data as you zoom in—at first the image is blurry, and then as streaming approaches 100%, it becomes clearer).

Topic 2.4   the panels sidebar

The sidebar contains information about locations and provides options for adding information to the screen image (Figure 2.5).

The topmost panel in the sidebar is the **Search panel** (Figure 2.5). There you can type a location (e.g., general name, specific address, latitude/longitude, etc.) in the text field and click the **Search** button to execute the search (and then fly directly to the location). To remove the list of search items, click the blue "X" that appears in the lower right corner of the **Search panel**.

The middle area is the **Places panel** (Figure 2.5). It stores locations that you create yourself or that you open using **File > Open** (e.g., *Geotours Workbook* files). Clicking the box next to a name will toggle between showing/hiding the feature. Double-clicking the icon (recommended) or name of an item within a folder will take you to that location. A toolbar at the bottom of the **Places panel** allows you to search for objects, adjust an object's opacity, and take a tour of the object(s). Search by clicking the **Magnifying Glass** icon (Figure 2.5) and then entering text into the adjacent text field. Folders will automatically expand to show you items that meet your search criteria. The arrows next to the text field at the bottom of the

> Why is it taking a long time to load my selected locations?
>
> Checking an entire folder will cause all of its contents to load into memory at once, slowing down the system. It is better to be selective about which features you toggle on at the same time. Also, having a large My Places folder will slow down opening Google Earth.

Places panel will take you to the previous (↑) or next (↓) instance that matches your search. Unfortunately, folders remain expanded and must be closed manually.

The **Places panel** has two main folders: **My Places** and **Temporary Places**. Items in the **My Places** folder are automatically saved each time you exit Google Earth, and the previous **My Places** folder is saved as a backup (or **File > Save > Save My Places**). To locate these files, look for myplaces.kml and myplaces.backup.kml on your system in the following directories (replace *username* with your login name):

- **Mac:** Macintosh HD/*username*/Library/Application Support/Google Earth/
- **Windows XP:** C:\Documents and Settings*username*\Application Data\Google/GoogleEarth\
- **Windows Vista:** C:\Users*username*\AppData\Roaming\Google\GoogleEarth\
- **Windows 7-10 (later Vista versions):** C:\Users*username*\AppData\LocalLow\Google\GoogleEarth\

If there is a corrupted **My Places** folder, you can delete myplaces.kml, copy/rename myplaces.backup.kml to myplaces.kml, and restart Google Earth. Items in the **Temporary Places** folder will not be automatically saved when you exit Google Earth. Instead, upon exiting Google Earth, a dialog will ask you if you want to move the unsaved items to your **My Places** folder. If you click **Save**, those files will be available in your **My Places** folder in the **Places** panel the next time you open Google Earth. Alternatively, you can drag items from the **Temporary Places** folder to the **My Places** folder (and vice versa) at any time.

The lowermost panel in the sidebar is the **Layers panel**, which shows content provided by Google Earth and associated collaborators (Figure 2.5). For example, when you click on "Borders and Labels" in the **Layers panel**, country and state boundaries as well as city names will appear to provide a visual reference frame.

Any of these three panels (**Search, Places,** and **Layers**) may be hidden by toggling the triangle next to their respective name (Figure 2.5).

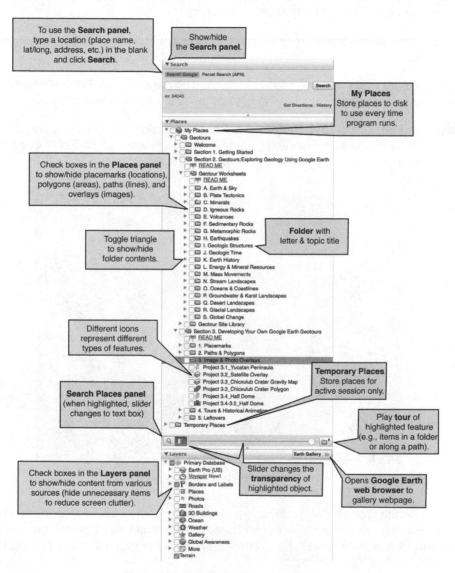

Figure 2.5: Google Earth panels sidebar.

Topic 2.5 | using the Search & Layers panels

The **Search** panel is your guide to traveling to all sorts of wonderful places: famous place names, specific addresses, even latitude and longitude coordinates. Once you click on the **Search** button to the right of the search text field, you will be whisked away to your specified locale (or offered a series of links from which to choose). Once you arrive at your location, you can explore the surrounding area by toggling various layers on and off in the **Layers** panel. The **Layers** panel is loaded with all kinds of neat content that we encourage you to explore (be forewarned—you're probably going to learn something).

In this project, we're going to search for sites using different techniques and then examine these sites using layers.

1. Type khufu pyramid, giza in the text blank of the **Search** panel. Press **Enter/Return** or click the **Search** button to fly to this area. Usually one of the first search links will fly you directly there.

2. In the **Layers** panel, turn on **3-D Buildings > Photorealistic**. Use the Navigation controls to change your perspective to zoom into the area and to look west at the Sphinx and the three pyramids of Giza, Egypt from an oblique angle (Figure 2.6).

Figure 2.6: The Great Pyramids of Giza, Egypt.

3. As you can see, the Google database has general place names for famous locations and features. It also can search for specific addresses. Why don't you type in *your address* and see if you can locate your home? For example, type your address in the following format: 1600 Pennsylvania Ave NW, Washington, DC 20500.

4. After typing in and flying to your address, go ahead and type in the address in step 3 and turn on **3D Buildings > Photorealistic** in the **Layers** panel and click the **Historical Imagery** clock icon. You should be treated to a detailed version of the White House, complete with trees and a flag (Figure 2.7). It is well worth your time to explore the National Mall (e.g., the Capitol Building, the Washington Monument, the Lincoln Memorial, etc.) with 3D Buildings turned on (and it will give you some good practice navigating within the Google Earth viewer). Turn off **Historical Imagery** when you are done.

Figure 2.7: A view from the South Lawn of the White House.

5. Turn on **Borders and Labels** in the **Layers** panel and type 36.092194, -112.118035 into the **Search** field (these numbers are "latitude, longitude"). You should fly west over the United States to Plateau Point in the Grand Canyon. If you didn't notice the various boundaries and place names, take a minute and zoom out to see them. Notice that the closer you are to the ground, the more detailed the information becomes.

6. Now turn on **More > Parks / Recreation Areas**. As you zoom back to Plateau Point, notice the wealth of information that is now available to you, including feature names, park boundaries, trails, etc. Rotate your perspective so that you can see the topography of the Grand Canyon expressed in Google Earth with the layer information draped over the 3-D surface (Figure 2.8).

> What do positive and negative latitudes and longitudes mean?
>
> Positive numbers indicate latitudes north of the equator and longitudes east of the prime meridian. Conversely, negative numbers indicate latitudes south of the equator and longitudes west of the prime meridian.

Note: In March, 2010, Google Earth removed the option to see more detailed labels (More > Geographic Features). Hopefully, these labels will be restored in response to user recommendations.

Figure 2.8: 3-D view of the South Rim of the Grand Canyon.

7. Lastly, turn everything off in the **Layers** panel except **Borders and Labels**. Turn on **Gallery > Earthquakes** and **Gallery > Volcanoes**. Now zoom into the Aleutian Islands in Alaska. With these data turned on, you can readily see that this is a plate tectonic boundary that has abundant recent seismicity and active volcanism associated with it (Figure 2.9).

Figure 2.9: Earthquake and volcanic data near the Aleutian trench.

Topic 2.6   miscellaneous hints & tips

General
- Please be patient as images are loading into memory as some placemarks will look incomplete until the images are completely loaded.
- It is easier to interact with Google Earth by using a mouse (preferably with a scroll wheel) than by using a trackpad.

Preferences (Mac)/Options (PC)
- Select **General > Show web results in external browser** as Google Earth's internal web browser sometimes freezes when attempting to return from a webpage.
- If the imagery has a third dimension and you want to emphasize or de-emphasize the elevations, select **3D View > Elevation Exaggeration**, which can be changed from 0.01 (nearly flat) to 3 (3x vertical exaggeration of elevations). A good general rule is to keep the elevation exaggeration at 1 (true elevation) or 1.5.
- If you don't like the auto-tilt feature as you zoom into the imagery, select **Navigation > Do not automatically tilt while zooming**.
- Set units using **3D View > Show Lat/Long** and **3D View > Units of Measurement**. A good general rule is to use **Decimal Degrees** with **Feet, Miles** and **Universal Transverse Mercator** with **Meters, Kilometers**, although you choices may certainly differ.
- To clear Google Earth's memory cache, first select **File > Server Sign Out** and then select **Cache > Delete cache file**. Now quit and restart the Google Earth application.

Panels
- If the imagery looks flat and lacks a third dimension, turn on **Terrain** in the **Layers panel** (Google Earth Pro only; terrain is always on in the latest version of Google Earth).
- Do not turn on all the folders at once because the program will load everything. This is true not only for the **Places panel**, but also for the **Layers panel** (especially 3D Buildings).
- You can make the Panels sidebar wider by clicking and dragging on the right side of the Panels sidebar (between it and the main Google Earth window).
- The **Search** and **Layers panels** can be hidden by clicking the triangle next to their name in order to see more of the **Places panel**.

Folders/Features (Places panel)
- Folders and placemarks can only be moved one at a time (i.e., multiple selections are not allowed). It may be useful to create a new folder first *(Add > New Folder)*, highlight it, and then create new placemarks within it.
- Any folder or feature can be saved as a KML or KMZ file. To do so, right-click on the folder or feature in the **Places** panel and choose **Save Place As**. In the Save dialog box, choose either the KML or KMZ format and indicate where you want to save the file.

> What is the difference between KML and KMZ files?
> A KML file is a plain text file that can be edited by any text editor, whereas a KMZ file is a compressed (zipped) file that is smaller than the KML file and cannot be read by a text editor.

- Recall from Topic 2.4 that items in the **My Places** folder are automatically saved when you exit Google Earth, whereas items in the **Temporary Places** folder are not. Fortunately, when you exit Google Earth, a dialog will ask you if you want to move the unsaved items from your **Temporary Places** folder to your **My Places** folder before closing. Alternatively, you can drag items from the **Temporary Places** folder to the **My Places** folder (and vice versa).

The Crab Nebula showcases a remnant of a spectacular supernova explosion observed by humans in 1054 C.E. Careful observations of the expanding shells of mostly helium and hydrogen gas over several years reveal a slow expansion of the nebula. Fusion of elements with atomic numbers above 26 likely occurred at the core of this massive explosion as lower elements fused into elements with larger nuclei.

Circle the letter next to the best answer for each question.

Worksheet Resources:
* **Google Earth** - Open the **2. Exploring Geology Using Geotours > A. Earth & Sky** folder.
  1. **Problem Materials** - **Check and double-click items associated with each problem** to travel to the appropriate location with the prescribed perspective/zoom.
  2. Earth & Sky Geotours Library - **Explore additional Geotours** in this folder to help answer problems.
* **Geoscience** - **Consult a textbook** and/or **Internet resources** to help answer some problems.

Select *Google Sky* by clicking 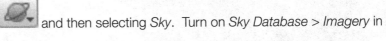 and then selecting *Sky*. Turn on *Sky Database > Imagery* in the *Layers* panel.

1. **Stellar Nursery - Pillars of Creation, Eagle Nebula.** The Pillars of Creation are large, dense masses of dust and interstellar gas (mostly molecular hydrogen) that rise from the stellar nursery of the Eagle Nebula (M16). Here, dense pockets of dust and gas collapse in on themselves to form young stars. The Pillars of Creation are located about 6500 light years from Earth, and the left-most pillar has a current length of approximately 4 light years (a **light year** is the distance light travels in one year...about 9.5 trillion km). What is the length of the left-most pillar (in km)?
   a. ~13.5 trillion km
   b. ~38 trillion km
   c. ~9.5 trillion km
   d. ~2.375 trillion km

2. **Main Sequence Stars - Tau Ceti.** If protostars accumulate sufficient mass as they develop, collapsing gas and dust may reach temperatures where hydrogen fuses into helium. Once hydrogen fusion occurs, stars become stable between the competing forces of fusion and gravity and are considered **main sequence stars**. Both our Sun and Tau Ceti are currently in this stage (both are considered **yellow dwarf stars**). For most stars, the main sequence stage lasts the longest *(with the amount of mass inversely proportional to how fast the star burns)*. Which type of star will have the longer main sequence stage, and therefore, "live" the longest?
   a. high-mass star (greater than 8x the mass of our Sun)
   b. intermediate- to low-mass stars (between 0.8x to 8x the mass of our Sun)

3. **Red Giant Stars - Aldebaran.** Eventually, the majority of hydrogen in the core of the main sequence stars is consumed, and the core collapses due to gravity. Very low-mass stars form **white dwarf stars**, whereas most other stars collapse and heat up until helium fusion begins. This fusion causes the star to expand outward several times larger than before, forming a **red giant star** (making the star cooler...Aldebaran is an example). When this happens to our Sun (in about 4.5 billion to 5 billion years), what will happen to Earth?
   a. Earth will be pushed by the expanding Sun to maintain its orbital distance from the Sun, allowing Earth to retain the same environmental conditions that it has now
   b. Earth will experience increased temperatures, but will retain its atmosphere and water
   c. Earth will lose its atmosphere and water and may/may not be completely vaporized by the Sun as it expands close to Earth's current orbit

4. **Planetary Nebula/White Dwarf Stars - Little Ghost Nebula.** Some red giant stars develop into **planetary nebulas** as their cores continue to contract, to increase in temperature, and to burn and vent the remaining gases into interstellar space. Eventually, the core collapses to the point where it is hot enough to ionize the vented gases, forming a relatively short-lived (~10,000-20,000 years) planetary nebula. The remaining core collapses into a **white dwarf star**. The vented materials from the planetary nebula play an important role in enriching the universe in elements with atomic weights less than 26 *(forming the basis for carbon-based life like ourselves!)*. Which type of star will form a planetary nebula?
   a. high-mass star (greater than 8x the mass of our Sun)
   b. intermediate- to low-mass stars (between 0.8x to 8x the mass of our Sun)

5. **Nebular Supernova - Crab Nebula.** Some red giant stars undergo a violent **supernova** explosion as opposed to forming planetary nebulas. Which type of star will form a supernova? *You may want to refer to your textbook or the Internet.*
   a. high-mass star (greater than 8x the mass of our Sun)
   b. intermediate- to low-mass stars (between 0.8x to 8x the mass of our Sun)

6. **Nebular Supernova - Crab Nebula.** The Crab Nebula represents a violent **supernova** explosion of a star (likely a third, fourth, or even later generation star). Heavy elements with atomic weights greater than 26 (and some with lesser atomic weights between oxygen and iron) were likely generated during this explosion. Which element probably formed in a supernova? *You may want to refer to your textbook or to a periodic table.*
   a. Gold (Au)
   b. Carbon (C)
   c. Hydrogen (H)
   d. Helium (He)

7. **Spiral Galaxy - M51.** This image is of a spiral galaxy (M51) that resembles what our Milky Way might look like if viewed from outside the galaxy. Note how the curved spiral arms develop around the more quickly rotating central cluster of stars. Looking at the spiral arms from this viewpoint, in what direction is this galaxy rotating?
   a. counterclockwise
   b. clockwise

Switch to *Google Moon* by clicking and then selecting *Moon*. Turn on *Layers > Moon Gallery > Apollo Missions > Apollo 11 & Apollo 16* and show *Layers > Moon Gallery > Historic Maps > Geologic Charts*.

8. **Impact Features - Moon.** *Apollo 11* touched down in the smooth, dark Mare Tranquillitatis (Sea of Tranquility), whereas *Apollo 16* landed in the Descartes lunar highlands. Which region contains impact features that are larger in size and that have a higher concentration density?
   a. Sea of Tranquility (Problem 8a placemark)
   b. Descartes lunar highlands (Problem 8b placemark)

9. **Impact Features - Moon.** Based on your answer to Question 8, which rock unit on the surface of the Moon is the youngest?
   a. pink (Problem 9a placemark)
   b. purple/brown/gray (Problem 9b placemark)

10. **Impact Features - Moon.** Turn off *Layers > Moon Gallery > Historic Maps > Geologic Charts*. Study the near side of the Moon (always faces toward Earth and has thinner crust) relative to the far side of the Moon (always faces away from Earth and has thicker crust). Which of the following best describes the nature of the dark maria?
   a. there is no difference in the nature of the maria between the two sides
   b. the near side has more maria (likely due to gravitational forces of Earth affecting crustal thicknesses and volcanic activity as the Moon formed)
   c. the far side has more maria (likely due to Earth shielding the near side from impacts)

Switch back to *Google Earth* by clicking [icon] and then selecting *Earth*. Now we'll visit impact features on Earth.

11. **Impact Features - Manicouagan Crater, Canada.** Use the *Ruler* tool to determine the present-day diameter of Manicouagan Crater between the placemarks.
   a. ~5 km
   b. ~1.2 km
   c. ~96 km
   d. ~75 km

12. **Impact Features - Meteor Crater, AZ.** Use the *Ruler* tool to determine the present-day diameter of Meteor Crater between the placemarks.
   a. ~5 km
   b. ~1.2 km
   c. ~96 km
   d. ~75 km

13. **Impact Features - Manicouagan Crater, Canada & Meteor Crater, AZ.** Assume that a 40-m-diameter meteorite created Meteor Crater. Although clearly an oversimplification, use a <u>simple ratio</u> between meteorite size and crater diameter to estimate the size of meteorite that might have created Manicouagan Crater (<u>use the crater diameters measured for Problems 11 & 12</u>).
   a. ~40 m
   b. ~170 m
   c. ~2500 m
   d. ~3500 m

14. **Impact Features - Parameters Influencing the Nature of Impact Features.** On your computer, go to the Earth Impact Effects Program website at: <u>http://www.lpl.arizona.edu/impacteffects/</u> . This site estimates the consequences of an impact as a function of various parameters, including the size, velocity, and composition of the meteorite. Enter the following:
   > *Distance from Impact: 1000 km*
   > *Projectile Diameter: Manicouagan's meteorite diameter (Problem 13, in m)*
   > *Projectile Density: <u>Trial 1</u>-ice (comet) and <u>Trial 2</u>-iron (some asteroids)*
   > *Impact Velocity: 20 km/s*
   > *Impact Angle: 45 degrees*
   > *Target Type: Crystalline Rock*

Perform Trials 1 & 2 to investigate the "impact" of changing projectile density. Which of the following statements is true?
   a. varying projectile density does not influence the final crater size
   b. icy comets would always melt before impacting the Earth and creating a crater
   c. an icy comet produces a larger final crater than an iron asteroid
   d. an icy comet produces a smaller final crater than an iron asteroid

<u>Just for Fun</u>...Use the *Ruler* tool and the crater diameters to see the consequence of what similar-sized impacts would be in the area in which you live!

Turn on the *Scaled Solar System* folder by clicking the check box next to it. Double-click the folder icon to zoom out to space to see the solar system scaled from <u>Los Angeles, CA (Sun)</u> to <u>New York, NY (Neptune)</u>. The placemarks in the folder contain numerical information about original and/or scaled parameters (e.g., radius, orbital distance, distance between objects).

15. **Solar System - Scaling.** The inner planets of our solar system can be classified as rocky, **terrestrial** (Earth-like) planets, whereas the outer planets are considered giant, gaseous **Jovian** (Jupiter-like) planets. Which of the following is correct relative to the scaled solar system?
    a. all of the terrestrial planets scale to reside entirely in California
    b. all of the terrestrial planets scale to reside entirely outside California
    c. the distance between previous objects increases for the Sun, Mercury, Venus, and Earth (i.e., distance between objects increases from Sun to Mercury, Mercury to Venus, and Venus to Earth)
    d. a & c
    e. b & c

16. **Solar System - Scaling.** Assume that you and a friend <u>both are capable of traveling a straight-line path from Earth to Jupiter</u>. You take a spacecraft that is capable of traveling an average speed of 70,811 km/hr (44,000 mi/hr) to the *actual* planet while your friend drives a vehicle at an average speed of 112.7 km/hr (70 mi/hr) along the red line of the *scaled* solar system. Which answer below is correct?
    a. 4.9 hr (car) vs 369.9 days (spacecraft)
    b. 7.9 hr (car) vs 595.2 days (spacecraft)
    c. 6.1 hr (car) vs 457.9 days (spacecraft)
    d. 4.9 hr (car) vs 515.3 days (spacecraft)

17. **Solar System - Scaling.** When you look at Neptune in a telescope, you are actually looking into the past as the light has to travel from Neptune to your eyes. If the speed of light is ~300,000 km/s, how far back into the past are you looking (or put another way, how long does it take light to travel from Neptune to your eyes on Earth)?
    a. ~14,501 hr
    b. ~4 hr
    c. ~242 hr
    d. ~1233 hr

18. **Solar System - Scaling.** Double-click the Problem 18 placemark to see the scaled size of Venus. Using this scaled model, what most closely approximates the area "footprint" of Venus?
    a. a small city
    b. a large factory/warehouse
    c. a city block
    d. a large backyard swimming pool

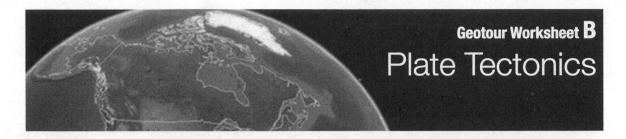

Plate Tectonics

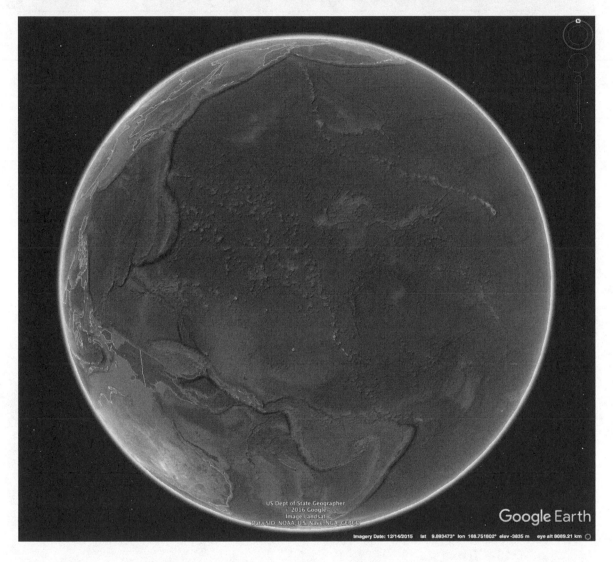

US Dept of State Geographer
2016 Google
Image Landsat
Data SIO, NOAA, U.S. Navy, NGA, GEBCO

Google Earth

Imagery Date: 12/14/2015   lat 9.893473° lon 168.751802° elev -3835 m   eye alt 8089.21 km

This view from outer space (09.901°N, 170.098°) captures the western margin of the Pacific Plate. With the water removed from the Pacific Ocean, details highlighted by the seafloor bathymetry are revealed. On the western (left) side of the image, the deep trenches where the Pacific Plate subducts and plunges back into the mantle are shown in dark hues from New Zealand in the south up to the Aleutian Islands in the north. The corresponding volcanic arc (e.g., Japan) on the overriding plate forms part of the "Ring of Fire" that surrounds the Pacific Plate. On the northeastern (right) side of the image, the trace of the Pacific Plate as it moved over a relatively stationary hot spot forms the continuous line of islands and seamounts that make up the Hawaiian Island–Emperor Seamount Chain. The prominent bend in the chain reflects a change in the direction of motion of the Pacific Plate.

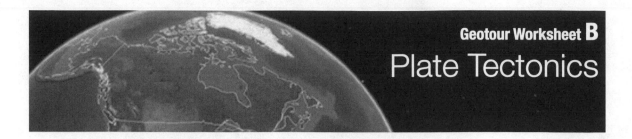

Geotour Worksheet **B**
Plate Tectonics

Circle the letter next to the best answer for each question.

Worksheet Resources:
- **Google Earth** - Open the **2. Exploring Geology Using Geotours > B. Plate Tectonics** folder.
  1. **Problem Materials** - **Check and double-click items associated with each problem** to travel to the appropriate location with the prescribed perspective/zoom.
  2. Plate Tectonics Geotours Library - **Explore additional Geotours** in this folder to help answer problems.
- **Geoscience** - **Consult a textbook** and/or **Internet resources** to help answer some problems.

Check the box next to the *Seafloor Age Map* folder to view the ages of the oceanic crust around the world. *Note that you can select the folder and use the transparency slider at the bottom of the Places panel to make the items in this folder semi-transparent.* Check and double-click the *Africa & South America* placemarks to fly to their positions on the opposite sides of the Atlantic Ocean. These placemarks represent **conjugate points** (locations on the opposite sides of an ocean that were once adjacent before seafloor spreading occurred). These points are located on the boundary between the **continental shelf** and the **continental slope**. Although the fit between the African and South American coastlines had been recognized for some time, Wegener showed that carefully matching continental shelves improves this fit.

1. **Seafloor Spreading - Atlantic Ocean.** Use the *Ruler* tool [ ] to determine how far these points have moved apart (in km). Select the *Path* tab on the *Ruler* tool and then create segments along the major fracture zone that offsets the colored ages of the ocean floor (round to the nearest 1000 km).
   - a. ~7400 km
   - b. ~2700 km
   - c. ~5300 km
   - d. ~3300 km
2. **Seafloor Spreading - Atlantic Ocean.** Using the *Seafloor Age Map*, about how many millions of years ago (mya) were these points once adjacent? *Note that the continental shelf and slope are not covered by the Seafloor Age Map as they are not composed of oceanic crust, so use the colored age band highlighted by the Problem 2 placemark (count the black isochron lines).*
   - a. ~115-120 mya
   - b. ~175-180 mya
   - c. ~55-60 mya
   - d. ~35-40 mya
3. **Seafloor Spreading - Atlantic Ocean.** Using the largest number of the range for your answer to Problem 2 and using the distance for your answer to Problem 1, calculate the average spreading rate for these points in km per million years.
   - a. ~58 km per million years
   - b. ~44 km per million years
   - c. ~83 km per million years
   - d. ~75 km per million years

4. **Seafloor Spreading - Atlantic Ocean.** Express your average spreading rate answer for Problem 3 in terms of cm per year. *Note: If the plates are moving apart symmetrically at the same rate (i.e., the color band widths are approximately equal), then 1/2 of this answer is the average rate at which the South American plate is moving west and the African plate is moving east.*
   a. 7.5 cm per year
   b. 5.8 cm per year
   c. 44.0 cm per year
   d. 4.4 cm per year

Just for Fun...Visit the region between Australia and Antarctica and see if you can determine conjugate points that were once joined. It is easier if the Seafloor Age Map is turned on!

Check the box next to the *Seafloor Age Map* folder to view the ages of the oceanic crust around the world. *Note that you can select the folder and use the transparency slider at the bottom of the Places panel to make the items in this folder semi-transparent.* Also, turn on the *Tectonic Plates* folder. Divergent = red, convergent = green, and transform = light blue.

5. **Divergent Boundaries - Mid-Atlantic Ridge vs East Pacific Rise.** The Problem 5 placemarks lie on the crests of mid-ocean ridges (**divergent boundaries**) in the Atlantic and Pacific Oceans, respectively. Double-click each placemark and use the width of the color bands representing seafloor ages to determine which divergent boundary is spreading at a faster rate.
   a. Problem 2a-Atlantic
   b. Problem 2a-Pacific
   c. both are spreading at approximately the same rate
6. **Divergent Boundary Spreading Rates - East Pacific Rise.** Some divergent boundaries spread at different rates. Use the widths of the color bands *(marked by the Problem 6 placemarks)* representing seafloor ages on either side of this mid-ocean ridge to determine which side is moving faster.
   a. east side
   b. west side
   c. both are spreading at approximately the same rate
7. **Divergent Boundary Triple Junction - Afar Triple Junction, Eastern Africa.** In the *Layers* panel, turn on *Borders and Labels* and also *Gallery > Volcanoes*, then double-click on the Problem 7 placemark. Divergent boundaries often begin as **triple junctions** composed of three rift "arms" at ~120° apart (which may or may not evolve into divergent boundaries). Knowing that rifts (red lines) generally open perpendicular to the rift (or sub-parallel to the light blue transform faults), which of the statements below about this area is <u>incorrect</u>?
   a. the NW-trending Red Sea is a linear sea that is opening in a NE-SW direction
   b. the ENE-trending Gulf of Aden is a linear sea that is opening in a NNE-SSW direction
   c. the volcanoes in Africa define a narrow linear rift that is opening in a NW-SE direction
   d. the volcanoes in Africa define a volcanic arc related to subduction off Africa's east coast

Just for Fun...Increase your vertical exaggeration to 3 (<u>On a Mac</u>: *Google Earth > Preferences > 3D View > Elevation Exaggeration* or <u>On a PC</u>: *Tools > Options > 3D View > Elevation Exaggeration*), zoom down to the line of volcanoes in Africa, and tilt your view. Fly over this line of volcanoes (known as the **East African Rift**) to view the landscape. Think about how this region may evolve similarly to Madagascar.

Check the boxes next to the *Tectonic Plates* and *Earthquakes* folders to view the tectonic plate boundaries (divergent = red, convergent = green, and transform = light blue) and to view selected 1986-2005 earthquakes color-coded by depth, respectively. Also, in the *Layers* panel, turn on *Borders and Labels* and *Gallery > Volcanoes*. Turn off the *Seafloor Age Map* folder.

8. **Convergent Boundary - Tonga Trench.** Check and double-click the Problem 8 placemark to fly to the Tonga Trench. Which of the following is <u>incorrect</u> (think about where volcanoes form in association with a subduction zone)?
   a. subducting plate is to the east and overriding plate is to the west
   b. subducting plate is to the west and overriding plate is to the east
   c. the line of volcanoes west of the Tonga Trench is the volcanic arc
   d. earthquakes generally become deeper to the west

9. **Convergent Boundary - Tonga Trench.** Subduction ultimately produces a volcanic arc on the overriding plate. What is the depth of the majority of earthquakes <u>directly</u> beneath the **volcanic arc** associated with the Tonga Trench?
   a. 0-50 km
   b. 51-100 km
   c. 101-200 km
   d. 201-400 km

10. **Convergent Boundary - Tonga Trench.** Use the *Ruler* tool to determine the **arc-trench gap** (in km) between the placemarks.
   a. ~245 km
   b. ~401 km
   c. ~295 km
   d. ~190 km

11. **Convergent Boundary - Nazca Trench.** Check and double-click the Problem 11 placemark to fly to South America. What is the depth of the majority of earthquakes <u>directly</u> beneath the **volcanic arc** associated with this subduction zone?
   a. 0-50 km
   b. 51-100 km
   c. 101-200 km
   d. 201-400 km

12. **Convergent Boundary - Nazca Trench.** Use the *Ruler* tool to determine the **arc-trench gap** (in km) between the placemarks.
   a. ~245 km
   b. ~401 km
   c. ~295 km
   d. ~190 km

13. **Convergent Boundary - Tonga Trench & Nazca Trench.** Assume that the earthquake depths define the Wadati-Benioff zone for the subducting slab. Compare your answers to Problems 9-12 and choose the statement that best describes your observations.
   a. subducting slabs must reach a depth of 51-100 km before they produce volcanic arcs
   b. subducting slabs must reach a depth of 101-200 km before they produce volcanic arcs
   c. arc-trench gaps are essentially the same distance for all subduction systems around the world
   d. the subducting slab has a steeper angle of descent for the larger arc-trench gap (Tonga) than for the smaller arc-trench gap (Nazca)

14. **Convergent Boundary - Tonga Trench & Nazca Trench.** Given your answers for the Tonga Trench (Problems 9 & 10) and for the Nazca Trench (Problems 11 & 12), calculate the angle of descent for the subducting slab for each subduction zone using the following formula *(note that this is just a rough approximation; use an average for the answers to Problems 9 & 11)*:
   angle of slab descent = tan⁻¹ (depth of slab beneath arc ÷ arc-trench gap).
   - a. Tonga = ~21° and Nazca = ~38°
   - b. Tonga = ~38° and Nazca = ~21°
   - c. Tonga = ~38° and Nazca = ~27°
   - d. Tonga = ~27° and Nazca = ~27°

Check the box next to the *Seafloor Age Map* folder to view the ages of the oceanic crust around the world. *Note that you can select the folder and use the transparency slider at the bottom of the Places panel to make the items in this folder semi-transparent.* Turn on the *Tectonic Plates* folder (divergent = red, convergent = green, and transform = light blue). Also, in the *Layers* panel, turn on *Borders and Labels* and *Gallery > Earthquakes*.

15. **Transform Boundary - Mid-Atlantic Ridge.** These placemarks point to a segment of either an active transform fault or an inactive fracture zone *(see the transform boundary section of your textbook)*. Which of the following statements is correct?
   - a. Problem 15-west and Problem 15-east are inactive fracture zones, and Problem 15-middle is an active transform fault
   - b. Problem 15-west and Problem 15-east are active transform faults, and Problem 15-middle is an inactive fracture zone
   - c. Problem 15-west and Problem 15-middle are inactive fracture zones, and Problem 15-east is an active transform fault
   - d. all three are along an active transform fault

16. **Transform Boundary - Mid-Atlantic Ridge.** Which direction are the placemarks for Problem 16-north and Problem 16-south moving, respectively? *Think about the process of rifting at the mid-ocean ridges.*
   - a. Problem 16-north is moving east, Problem 16-south is moving west
   - b. Problem 16-north is moving west, Problem 16-south is moving east
   - c. both are moving east
   - d. both are moving west

17. **Transform Boundary - Mid-Atlantic Ridge.** Which direction are the placemarks for Problem 17-north and Problem 17-south moving, respectively? *Think about the process of rifting at the mid-ocean ridges.*
   - a. Problem 17-north is moving east, Problem 17-south is moving west
   - b. Problem 17-north is moving west, Problem 17-south is moving east
   - c. both are moving east
   - d. both are moving west

18. **Transform Boundary - San Andreas Fault, CA.** Check and double-click the Problem 18 folder to see a drywash (blue line) that was offset by the San Andreas Fault. What type of transform fault is the San Andreas Fault? *Imagine walking toward the fault zone along the offset feature, and think about which direction you would look to see the other piece of the offset feature.*
   - a. right-lateral *(at the fault, you turn right to find the offset wash segment)*
   - b. left-lateral *(at the fault, you turn left to find the offset wash segment)*

Check the box next to the *Seafloor Age Map* folder to view the ages of the oceanic crust around the world. *Note that you can select the folder and use the transparency slider at the bottom of the Places panel to make the items in this folder semi-transparent.* Turn on the *Tectonic Plates* folder (divergent = red, convergent = green, and transform = light blue). Also, in the *Layers* panel, turn on *Borders and Labels* and *Gallery > Volcanoes*.

19. **Oceanic Hot Spots - Hawaiian Island–Emperor Seamount Chain.** Make the *Seafloor Age Map* semi-transparent, and turn on the *Hawaiian Island–Emperor Seamount Chain* folder to show their age of formation in millions of years (Ma). Use the *Ruler* tool to measure the distance between Midway Atoll and Kilauea (cm; Problem 19 placemarks) and calculate the average velocity of the Pacific Plate in cm per year.
    a. ~1.9 cm per year
    b. ~4.2 cm per year
    c. ~8.8 cm per year
    d. ~15 cm per year

20. **Oceanic Hot Spots - Hawaiian Island–Emperor Seamount Chain.** What is the approximate age of the bend in the Hawaiian Island–Emperor Seamount hot-spot chain (should you use the age of the volcanic features or the age of the seafloor)?
    a. ~125 Ma
    b. ~150 Ma
    c. ~32 Ma
    d. ~43 Ma

Check the box next to the *Yellowstone Hot-Spot Trail* folder to view the locations and ages of the main calderas that formed in association with the Yellowstone continental hot spot over the past 16.5 Ma. *Note that you can select the folder and use the transparency slider at the bottom of the Places panel to make the items in this folder semi-transparent.*

21. **Continental Hot Spots - Yellowstone Hot-Spot Trail.** Use the *Path* tab on the *Ruler* tool to measure the distance between the Problem 21 placemarks (cm) along the hot-spot trail. Calculate the average velocity of the North American Plate in cm per year using 16.5 Ma and 0.63 Ma for the two placemarks.
    a. ~1.9 cm per year
    b. ~4.2 cm per year
    c. ~8.8 cm per year
    d. ~15 cm per year

22. **Continental Hot Spots - Hawaiian Island–Emperor Seamount Chain & Yellowstone Hot-Spot Trail.** Which of the following best describes the present-day direction and rate of motion of both the Pacific Plate and the North American Plate (refer to Problems 19 & 21)? *Think about the analogy of a piece of paper passing over a candle.*
    a. Pacific Plate moves SE at about half the rate of the NE-moving North American Plate
    b. Pacific Plate moves NW at about the same rate of the NE-moving North American Plate
    c. Pacific Plate moves SE at about twice the rate of the SW-moving North American Plate
    d. Pacific Plate moves NW at about twice the rate of the SW-moving North American Plate

This remote area of the frigid Canadian Arctic contains very precious minerals that warm our hearts...diamonds! From this perspective, the open mine pits of the Ekati Diamond Mine (64.722°N, 110.615°W) northeast of Yellowknife, Canada form deep scars in the tundra landscape as the diamond-containing ore is quarried from kimberlite volcanic pipes. These pipes are thought to have originated deep in the mantle where the diamonds formed under great pressures and then were rapidly (and perhaps violently) emplaced along sub-vertical conduits around 45 to 60 millions of years ago.

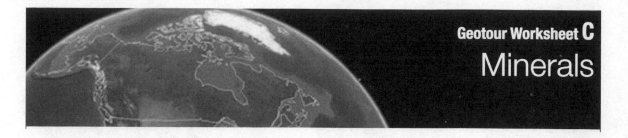

Circle the letter next to the best answer for each question.

Worksheet Resources:
- **Google Earth** - Open the **2. Exploring Geology Using Geotours > C. Minerals** folder.
  1. **Problem Materials** - **Check and double-click items associated with each problem** to travel to the appropriate location with the prescribed perspective/zoom.
  2. **Minerals Geotours Library** - **Explore additional Geotours** in this folder to help answer problems.
- **Geoscience** - **Consult a textbook** and/or **Internet resources** to help answer some problems.

Many of today's high-tech devices (such as cell phones, hybrid cars, wind turbines, etc.) depend on minerals that contain **rare earth elements** (**REEs**; 17 elements in the periodic table including the 15 lanthanides plus yttrium and scandium). These elements aren't actually rare, but must be found in concentrated quantities that can be economically extracted. In addition, refining the mineral ores that contain REEs poses significant environmental challenges.

1. **REE - Baotou, China.** The region around Baotou, China currently provides more than 90% of the world's supply of REEs despite China having less than 40% of the world's proven reserves *(reflecting, in part, regulation regarding environmental issues pertaining to the refining process)*. The Problem 1 placemark shows a large artificial lake created to hold the slurry of liquid and solid waste products from the refining process (including radioactive thorium and uranium). Use the *Polygon* tab on the *Ruler* tool to measure the area (acres) of the tailings pile and lake (i.e., trace the boundary highlighted by the Problem 1 placemark).
   a. ~2000 acres
   b. ~40 acres
   c. ~640 acres

2. **REE - Baotou, China.** In general, groundwater in this region flows from the high mountainous region north of Baotou toward the south. Assume that you are a geologist for the REE mining company, what potential environmental concerns might you have for this region?
   a. contamination of groundwater-based drinking water (especially S of the artificial lake)
   b. decreased crop production due to irrigation with contaminated water and due to contaminated soil adjacent to the area
   c. contamination of the Yellow River (S of city), which potentially might affect drinking water for downstream inhabitants
   d. all of the above

3. **REE - Mountain Pass, CA & Baotou, China.** One of the highest concentrations of REEs discovered to date in the United States is located in Mountain Pass, CA. Here, Molycorp is developing techniques to reduce the environmental challenges of REE processing by recycling waste liquids back into the refining process and encasing waste solids in cement within a nearby lined disposal site. Which of the following best compares these two REE mining areas?
   a. Mountain Pass has more potential than Baotou to affect humans because of Mountain Pass's close proximity to human homes and farms
   b. Mountain Pass has less potential than Baotou to contaminate groundwater because of Mountain Pass's waste disposal process
   c. Baotou has less potential than Mountain Pass to contaminate groundwater because of Baotou's drier climate and sparse vegetation
   d. Baotou has less potential than Mountain Pass to contaminate offsite areas because Baotou has fewer surface streams than Mountain Pass

In addition to REEs, many of the technologies that we use every day (such as cell phones and laptop computers) require minerals like columbite-tantalite (source of tantalum), cassiterite (source of tin), wolframite (source of tungsten), and native gold. Collectively, these four substances are referred to as **3TG**, and they form the basis for the term **"conflict mineral."** Conflict minerals are minerals (or their derivatives) used to finance armed conflict in the Democratic Republic of the Congo (**DRC**) and adjoining countries and often are acquired under conditions of human rights abuses. Toggle the *Conflict Minerals > Countries* folder on/off to view the relevant countries. Turn on the *Conflict Minerals > Tectonic Plates* folder to show the plate tectonic boundaries (divergent = red, convergent = green, and transform = light blue).

4. **Conflict Minerals - DRC, Africa.** Turn on the *Conflict Minerals > Mineral Resources* folder to show the distribution of mineral resources in the DRC. Where are most of the 3TG conflict minerals in the DRC located?
   a. eastern part of DRC
   b. western part of DRC
5. **Conflict Minerals - DRC, Africa.** How does the occurrence of conflict minerals spatially relate to plate tectonic boundaries (divergent = red, convergent = green, and transform = light blue)?
   a. conflict minerals seem to be concentrated on the western margin of a divergent continental rift (called the East African Rift)
   b. conflict minerals seem to be concentrated on the western margin of a volcanic arc formed by convergent subduction (called the East African Subduction Zone)
   c. conflict minerals seem to be concentrated on the western margin of a major transform fault (called the East African Fault)
6. **Conflict Minerals - DRC, Africa.** Where are most of the conflict diamonds in the DRC located?
   a. north-central part of DRC
   b. south-central part of DRC

Minerals have a wide variety of **physical properties** that make them useful in today's society. Below, we will visit areas and learn about the uses and/or physical properties of various common minerals.

7. **Gypsum - Naica, Mexico.** The Problem 7 placemark takes you to Cave of the Crystals in Naica, Mexico. The cave is home to the largest **selenite** (a variety of gypsum) crystals in the world. **Gypsum** ($CaSO_4 \cdot 2H_2O$) is a representative member of the **sulfate** mineral class where metal cations bond with the sulfate $(SO_4)^{2-}$ anion. These crystals formed over hundreds of thousands of years as calcium from nearby limestones interacted with groundwater rich in sulfide ions from neighboring ores and heated by a subterranean magma chamber. Gypsum (which can also be a sedimentary rock) has many uses for society. Which of the following is not a common use of gypsum?
   a. wallboard/drywall for homes
   b. a primary ingredient for toothpaste
   c. surgical splints and casts
   d. pH reducing agent for acidic soils

8. **Galena - Coeur d'Alene Mining District/Silver Valley, ID.** The Coeur d'Alene Mining District in Silver Valley, ID is known for its large lead deposits of the mineral **galena** (PbS). Galena is one of the few common minerals with a **metallic luster** that has a distinctive cleavage (**cleavage** is how a mineral breaks or parts along preexisting planes of weakness). Cleavage is described by the number of unique directions of cleavage planes and the angle between those planes. Which of the following best describes galena's cleavage (see animation in placemark)?
   a. three directions at ~90°
   b. three directions not at ~90°
   c. two directions at ~90°
   d. two directions not at ~90°

9. **Specular Hematite - Ishpeming, MI.** Many minerals with metallic lusters produce a prominent **streak** (aka "true color") when rubbed on unglazed porcelain. For example, galena's streak is dark gray (very much like its appearance); however, that is not always the case. Turn on the zoomable photo for Problem 9. Which of the following best describes the streak for metallic gray **specular hematite** (Fe_2O_3)?
   a. red-brown
   b. yellow
   c. gray
   d. black

10. **Calcite - Eskifjordur, Iceland.** This remote area of Iceland is the type locality for the Iceland Spar variety of **calcite** ($CaCO_3$; **type locality** is the place where a mineral or rock best exemplifies its standard description, often where it was first discovered). Iceland Spar calcite is exceptionally clear and optically transparent, whereas other types of calcite can have a wide range of colors and translucencies (see photo in the Problem 10 placemark). Given this, which of the following is not a diagnostic property for calcite?
   a. cleavage at three directions not at ~90°
   b. hardness of 3
   c. color
   d. effervesces (forms CO_2 bubbles) with hydrochloric acid

11. **Fluorite - Cave-in-Rock, IL.** The Problem 11 placemark takes you to the Illinois Fluorite District near Cave-in-Rock, IL, which once was responsible for more than 50% of total US fluorite production (now all fluorite is imported into the United States). **Fluorite** (CaF_2) forms as cubic **crystals**, but when broken, it exhibits cleavage that forms triangular luster faces that look like two pyramids stacked on top of each other. How many <u>directions of cleavage</u> does fluorite exhibit (see animation in placemark)?
    a. eight
    b. four
    c. two
    d. three

As we have seen in this Geotour Worksheet, mining and processing of rocks and minerals often leads to unintended environmental consequences, resulting in the need to reclaim and remediate mining areas to reduce these hazardous effects. Sometimes, however, even the best attempts at remediating former mining sites can face challenges.

12. **Pyrite Weathering - Green Valley Strip Mine, West Terre Haute, IN.** Double-click the folder for Problem 12 to fly to a former coal strip mine in western Indiana that has been reclaimed. Minerals like **pyrite** (FeS_2) are commonly associated with coal and have been brought to the surface during the mining process where they can react and chemically weather more readily. Specifically, sulfuric acid and iron oxide (orange staining) are produced. Within the Problem 12 folder, there are two folders *(March 2005 and November 2013)*. Check and double-click all of the placemarks in each folder and then take the *Little Sugar Creek* flyover in each folder *(March 2005 first and then November 2013; for the flyovers, set Preferences (Mac)/Options (PC) > Touring > Camera Tilt Angle = 65°, Camera Range = 1000 m, and Speed = 900)*. Which of the following is <u>not correct</u>?
    a. the amount of acidic, iron-rich water flowing down Little Sugar Creek into the Wabash River has increased from 2005 to 2013
    b. the amount of acidic, iron-rich water leaching from the reclaimed mine area into nearby ponds has decreased from 2005 to 2013
    c. the remediation process has not been 100% successful in containing waste runoff at the mine site in either year

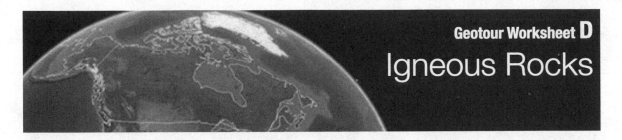

Most anywhere you look in this image of Yosemite Valley (37.737°N, 119.548°W) and the surrounding environs, you will see a landscape dominated by igneous rock that originated deep underground (perhaps ~20-25 km) as a massive pluton related to subduction along the western margin of North America around 80 to 120 millions of years ago. The granites and granodiorites that compose this batholithic intrusion (today known collectively as the Sierra Nevada mountains) cooled slowly to form coarse grains of quartz, feldspar, and mica minerals. The rocks were uplifted and exposed during Basin and Range extension and later sculpted by glaciers to form the beautiful features that you see today.

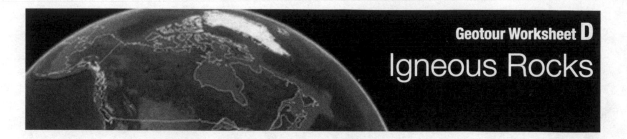

Igneous Rocks

Circle the letter next to the best answer for each question.

<u>**Worksheet Resources:**</u>
- **Google Earth** - Open the **2. Exploring Geology Using Geotours > D. Igneous Rocks** folder.
  1. <u>**Problem Materials**</u> - **Check and double-click items associated with each problem** to travel to the appropriate location with the prescribed perspective/zoom.
  2. <u>Igneous Rocks Geotours Library</u> - **Explore additional Geotours** in this folder to help answer problems.
- **Geoscience** - **Consult a textbook** and/or **Internet resources** to help answer some problems.

---

1. **Batholiths - Yosemite National Park, CA.** Yosemite National Park provides spectacular glaciated exposures of the Sierra Nevada batholith. The **batholith** is composed of many separate plutons, which intruded into the area as part of a Mesozoic-age subduction system. The Problem 1 placemark flies you to the spectacular Half Dome in Yosemite National Park. Given that it is part of an <u>intrusive</u> batholith, what type of rock texture would you expect to see at Half Dome?
   - a. fine-grained
   - b. coarse-grained
   - c. glassy
   - d. fragmental

2. **Batholiths - Yosemite National Park, CA.** Visitors commonly note the rounded appearance of the back side of Half Dome. It was not polished by glaciers, but rather fractured off in sheets much like how one might peel layers from an onion. This process occurs because these rocks formed at ~20-25 km depth, but have experienced a removal of confining pressure because they are now exposed at the surface *(think of how a sponge expands once you remove your hands pressing down on it)*. What is this process called? *(Hint: Look these words up in a glossary.)*
   - a. stoping
   - b. assimilation
   - c. fractional crystallization
   - d. exfoliation

3. **Batholiths - Sierra Nevada Mountains, CA.** Turn on the *North America Batholiths* overlay. You will see the exposed batholiths of the western United States in a red-brown color. Using the transparency slider at the bottom of the Places panel, make the overlay semi-transparent. To get a sense of the size of the Sierra Nevada batholith, determine its length between the Problem 3 placemarks using the *Ruler* tool (in km).
   - a. 100-130 km
   - b. 300-340 km
   - c. 700-720 km
   - d. 1000-1030 km

4. **Laccoliths - Henry Mountains, UT.** Turn on the red polygon for the *Henry Mountains Laccolith Complex* (leave the *North America Batholiths* overlay on). This complex consists of many blister-like intrusions. Note the size difference between this intrusion and the Sierra Nevada batholith. Estimate how much smaller in length the Henry Mountains laccolith complex is relative to the batholith.
   a. ~1/2
   b. ~1/4
   c. ~1/6
   d. ~1/17

5. **Tabular Intrusions - Mt. Hillers, UT.** Turn off the *North America Batholiths* overlay and the *Henry Mountains Laccolith Complex* polygon, and then turn on and double-click the Problem 5 placemark to zoom down to the south flank of the Mt. Hillers laccolith. The laccolith is composed of gray andesite that has intruded into Mesozoic-age sedimentary rocks and upturned them on the flanks of the intrusion *(recall that **laccoliths** are concordant intrusions that cause the overlying sedimentary layers to fold up in a blister or mushroom-shaped pattern)*. The sedimentary rocks over the intrusion have been eroded. The Problem 5 placemark highlights a tabular intrusion from the main laccolith that is parallel/concordant to the upturned sedimentary layers. What kind of intrusion is this?
   a. sill *(a tabular intrusion that inserts between sedimentary layers)*
   b. dike *(a tabular intrusion that cuts across sedimentary layers)*

6. **Tabular Intrusions - Spanish Peaks, CO.** Recalling that a **dike** is a tabular igneous intrusion that is discordant and cuts across preexisting layering, fly to the Spanish Peaks, CO area where some prominent volcanic centers are visible. Which of the following statements is incorrect?
   a. igneous rock composing the dike is more resistant to erosion than the surrounding rock
   b. there are no similar dikes in the surrounding area
   c. the dike is very linear, implying that magma likely filled a crack
   d. the dike's magma source is likely related to the Spanish Peaks volcanic centers

7. **Tabular Intrusions - Shiprock, NM.** The Problem 7 placemarks highlight dikes filling radial fractures from the volcanic neck at Shiprock. Which of the following is the most reasonable explanation for this pattern?
   a. they filled fractures that existed in the area before the volcano developed
   b. the volcanic magma chamber created pressures that fractured the rocks in a radial pattern
   c. this fracture pattern is accidental and random
   d. the fracture pattern is controlled by the nearly horizontal sedimentary rock layers

Just for Fun...Check and double-click on the polygon for *Stone Mt., GA* to fly to that location. Use the *Ruler* tool to measure the long axis of the stock that composes Stone Mt. Clearly, this stock is much smaller than the Sierra Nevada batholith represented on the *North America Batholiths* overlay (recall that **stocks** often are the exposed tips of larger subsurface plutons like batholiths). In the future, given enough additional time and erosion, this area may be considered a batholith as well!

On May 18, 1980 at 8:32 am, Mt. Saint Helens (46.199°N, 122.187°W), a composite cone/stratovolcano that is part of the volcanic arc that makes up the Cascade Range, erupted and markedly changed the landscape that you see before you in southwest Washington State. The initial blast erupted laterally out the north flank of the volcano, devastating an area of ~600 km² (~230 mi²) and removing ~915 m (~3000 ft) from the top of the mountain. Lahars (mudslides of ash and other pyroclastic material) triggered by melting snow flowed down the flanks of the volcano to travel miles along streams draining the area. Mount Hood, another active volcano in the Cascade Range, is seen in the distance to the southeast.

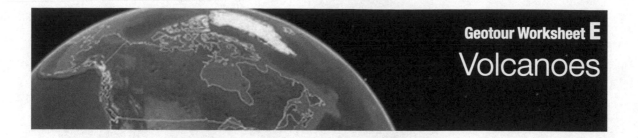

Circle the letter next to the best answer for each question.

Worksheet Resources:
- **Google Earth** - Open the **2. Exploring Geology Using Geotours > E. Volcanoes** folder.
  1. **Problem Materials** - **Check and double-click items associated with each problem** to travel to the appropriate location with the prescribed perspective/zoom.
  2. **Volcanoes Geotours Library** - **Explore additional Geotours** in this folder to help answer problems.
- **Geoscience** - **Consult a textbook** and/or **Internet resources** to help answer some problems.

Shield volcanoes have a characteristic shape like a concave-down warrior's shield, in part because they are predominantly composed of low-viscosity basaltic lava.

1. **Shield Volcanoes - Mauna Loa, HI.** Fly to Mauna Loa on the Big Island of Hawaii in the Hawaiian Islands, and then turn on the *Mauna Loa Contour Map* overlay. The brown lines are **contours** (lines of equal elevations). Use the *Ruler* tool to determine the **horizontal distance** (ft) between the Problem 1 placemarks and then subtract the elevations (ft) for each placemark from the contour map to determine the **relief** (vertical distance). Find the **slope angle** of Mauna Loa using the following formula (*this value is representative for many shield volcanoes*):
   slope angle = tan^{-1} (relief ÷ horizontal distance)
   - a. 5°-10°
   - b. 15°-20°
   - c. 25°-35°
   - d. 35°-45°
2. **Shield Volcanoes - Kilauea, HI.** The basaltic lava flows weather according to their age (dark/gray is younger than light/brown). Place the three lava flows indicated by the Problem 2 placemarks in chronologic order from youngest to oldest.
   - a. Problem 2a, 2b, and 2c
   - b. Problem 2b, 2c, and 2a
   - c. Problem 2c, 2b, and 2a
   - d. Problem 2b, 2a, and 2c
3. **Shield Volcanoes - Kilauea, HI.** The Problem 3 placemark highlights where the fluid basaltic lava flows have traveled from Kilauea's eruptive vents over a steep cliff on their way to the ocean. Such a cliff is called a **pali**. Geologically, what does a pali represent (*check and double-click the Hawaiian Island Landslides overlay*)?
   - a. the uppermost part of a landslide where the land has detached and moved to the SE
   - b. an edge of a previous lava flow
   - c. the marginal flank of an offshore caldera
   - d. the erosional remnant of a former river bank
4. **Shield Volcanoes - Kilauea, HI.** Even from this distance, you can see the fluid, ropey flow patterns of this distinctive type of basaltic lava flow. What is the term for this kind of lava flow?
   - a. pahoehoe
   - b. a'a'

5. **Shield Volcanoes - Mauna Kea, HI.** Along the road to the summit of Mauna Kea, a series of small parasitic volcanoes dot the landscape. These types of volcanoes often erupt during the latter stages of activity for the larger volcano and are composed of mostly pyroclastic material. What type of volcanoes are they?
   a. shield
   b. composite cone
   c. stratovolcano
   d. cinder cone

Switch to *Google Mars* by clicking and then selecting *Mars*. Earth isn't the only home to volcanoes in the solar system!

6. **Shield Volcanoes - Olympus Mons, Mars.** Turn on the *Olympus Mons, Mars* polygon (make sure that *Layers > Global Maps > Visible Imagery* is on) to see the area encompassing the largest known shield volcano in the solar system. Click *RMB > Get Info (Mac) or Options (PC) > Measurements* to see the approximate area (sq km) of Olympus Mons. Which state is closest to the area of Olympus Mons (see https://simple.wikipedia.org/wiki/List_of_U.S._states_by_area for state areas)?
   a. Wyoming
   b. Rhode Island
   c. Indiana
   d. Olympus Mons is much smaller than any state

Switch back to *Google Earth* by clicking and then selecting *Earth*. Now we'll visit other types of volcanoes on Earth.

Stratovolcanoes (or **composite cones**) have the characteristic symmetrical shape most commonly associated with volcanoes. They are typically composed of layers of andesitic lava and pyroclastic materials (although they sometimes can be very basaltic too). Because of their composition, these volcanoes tend to be very explosive.

7. **Stratovolcanoes - Mt. Saint Helens, WA.** Fly to Mt. Saint Helens in the Cascade Range of western Washington, and then turn on the *Mt. Saint Helens Contour Map* overlay. The brown lines are **contours** (lines of equal elevations). Use the *Ruler* tool to determine the **horizontal distance** (ft) between the Problem 7 placemarks and then subtract the elevations (ft) for each placemark from the contour map to determine the **relief** (vertical distance). Find the **slope angle** of Mt. Saint Helens using the following formula *(this value is representative for many composite cone volcanoes)*: slope angle = tan^{-1} (relief ÷ horizontal distance)
   a. 5°-10°
   b. 15°-20°
   c. 25°-35°
   d. 35°-45°
8. **Stratovolcanoes - Mt. Saint Helens, WA.** Turn on the *Mt. Saint Helens Volcanic Features* overlay. Use the *Ruler* tool to determine the greatest distance (in km) affected by the lateral blast of the May 18, 1980 eruption (measure between the Problem 8 placemarks).
   a. 98-101 km
   b. 16-21 km
   c. 40-44 km
   d. 22-28 km

9. **Stratovolcanoes - Mt. Etna, Sicily, Italy.** Fly to the town of Bronte near Mt. Etna, a large mountain on the island of Sicily. You are a geologic consultant for a company that is considering building a factory in Bronte. Based on your previous work on Mt. Saint Helens and on observations that you make as you fly over the region, which of the following is the answer that you would give the company regarding whether to build here or not?
   a. no, Bronte is about 15 km from an active composite cone volcano
   b. no, a previous lava flow has covered the eastern parts of town
   c. no, because of both of the previous reasons
   d. yes, the weather here is great

10. **Stratovolcanoes - Mt. Vesuvius, Italy.** Fly to Mt. Vesuvius in Italy and turn on the *Pompeii Pyroclastic Flow* tour *(run the tour twice to let the imagery load)*. Use the *Ruler* tool to determine the distance (in km) from Mt. Vesuvius to the excavated ruins of the city of Pompeii (which was buried by volcanic ash and pyroclastic material from Mt. Vesuvius during the eruption of 79 C.E.) If a **nuée ardent** (pyroclastic surge) accompanies the next eruption of Mt. Vesuvius and it travels at 300 km/hr, how many <u>minutes</u> would it take to reach the site of ancient Pompeii?
   a. ~2 min
   b. ~0.2 min
   c. ~5 min
   d. ~7 min

11. **Stratovolcanoes - Crater Lake National Park, OR.** Fly to Crater Lake, OR, and turn on the *Crater Lake, OR Map* overlay. The composite cone Mt. Mazama explosively erupted ~7000 years ago. Not only was most of the cone blown away, but a large depression was created as the material collapsed into the evacuated magma chamber. What is this depression called? *Note the parasitic* **cinder cone** *(Wizard Island) that has developed.*
   a. crater *(Crater Lake is named appropriately)*
   b. caldera *(Crater Lake is actually incorrectly named)*
   c. chamber basin *(Crater Lake is actually incorrectly named)*
   d. evacuation basin *(Crater Lake is actually incorrectly named)*

Cinder cones are the smallest volcanoes and often are found parasitic on larger volcanoes. They typically erupt pyroclastic materials, but they also commonly have small lava flows that emit from the base of the cone during the latter parts of their short-lived eruptive "lives".

12. **Cinder Cones - S P Mountain, AZ.** Fly to S P Mountain, AZ, and turn on the *S P Mountain Contour Map* overlay. The brown lines are **contours** (lines of equal elevations). Use the *Ruler* tool to determine the **horizontal distance** (ft) between the Problem 12 placemarks and then subtract the elevations (ft) for each placemark from the contour map to determine the **relief** (vertical distance). Find the **slope angle** of S P Mountain using the following formula (this value is representative for many cinder cone volcanoes): <u>slope angle = $\tan^{-1}$ (relief ÷ horizontal distance)</u>
   a. 5°-10°
   b. 15°-20°
   c. 25°-35°
   d. 35°-45°

13. **Cinder Cones - Menan Buttes, ID.** These cinder cones are dominated by pyroclastic material and are very asymmetric. What causes the asymmetry *(think about the type of material and the direction of asymmetry)*?
   a. the cones have been reworked by the nearby river
   b. cinder cones are composed of pyroclastic material, which can be blown by winds
   c. this is just a typical erosion pattern of cinder cones
   d. landsliding of material

14. **Cinder Cones - Sunset Crater, AZ.** To get a sense of scale for cinder cones, use the *Ruler* tool to determine Sunset Crater's width (in km) between the Problem 14 placemarks.

     a. 10-15 km

     b. 5-10 km

     c. 15-20 km

     d. 1-3 km

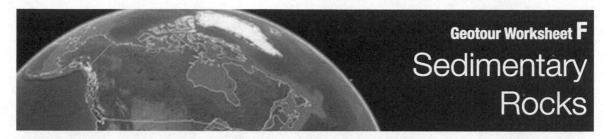

Geotour Worksheet **F**

Sedimentary Rocks

The Grand Canyon (36.100°N, 112.106°W) exposes rocks that record ~2 billion years of Earth's geologic history. The layered sedimentary rocks reveal the rise and fall of sea level over an immense expanse of time, capturing a snapshot of not only what climatic and depositional conditions existed at that time, but also what life-forms were present before the sediment and limestone mud hardened into sedimentary rock. In addition, these rocks retain information about the geologic processes that acted upon these layers before, during, and after lithification.

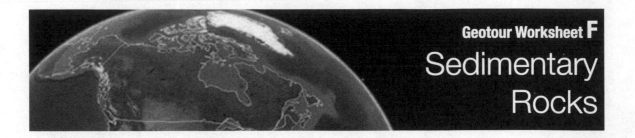

Geotour Worksheet F
Sedimentary Rocks

Circle the letter next to the best answer for each question.

Worksheet Resources:
- **Google Earth** - Open the **2. Exploring Geology Using Geotours > F. Sedimentary Rocks** folder.
  1. **Problem Materials** - **Check and double-click items associated with each problem** to travel to the appropriate location with the prescribed perspective/zoom.
  2. **Sedimentary Rocks Geotours Library** - **Explore additional Geotours** in this folder to help answer problems.
- **Geoscience** - **Consult a textbook** and/or **Internet resources** to help answer some problems.

The Grand Canyon is a marvelous place to study geology. Here, **weathering** and **erosion** by the Colorado River have revealed an immense span of time as recorded in the rock record. Turn on the *Grand Canyon Geologic Map* overlay and make it semi-transparent *(Note: This overlay is 4 MB and takes time to be downloaded into Google Earth's memory cache. Slight mismatches exist because the map is draped over topography.)* In this worksheet, we're going to focus on the **cover sequence** of sedimentary rocks that were deposited on top of the older metamorphic and igneous **basement** rocks.

1. **Erosion - Grand Canyon National Park (NP), AZ.** In order to get a sense of **horizontal scale** of this erosional exposure of sedimentary rocks, use the *Ruler* tool to determine the distance (in km) between the North and South Rims between the Problem 1 placemarks.
   a. 14-16 km
   b. 4-7 km
   c. 22-25 km
   d. 30-33 km

2. **Erosion - Grand Canyon NP, AZ.** To get a sense of erosional **relief** in this same area, hover the hand cursor over each Problem 2 placemark and find the relief (m) by subtracting the elevations shown at the bottom of the screen in the *Information Bar*.
   a. 2310-2340 m
   b. 1400-1425 m
   c. 975-1000 m
   d. 3020-3050 m

3. **Differential Weathering - Grand Canyon NP, AZ.** To identify the boundary between the sedimentary **cover** rocks (above) and the igneous and metamorphic **basement** rocks (below), turn on Problem 3 placemarks to delineate the steep and narrow Inner Gorge adjacent to the Colorado River. Which of the following statements is incorrect?
   a. crystalline metamorphic & igneous basement rocks are located in the Inner Gorge
   b. the Inner Gorge is ~0.5-0.7 km wide here
   c. sedimentary cover rocks all weather to form uniformly similar slopes
   d. sedimentary cover rocks more easily weather to form the wider parts of the canyon

4. **Differential Weathering - Grand Canyon NP, AZ.** Locate the Redwall Limestone (Mr) highlighted by the Problem 4 placemark. What kind of slope does it form?
   a. steep cliff
   b. gentle slope

5. **Differential Weathering - Grand Canyon NP, AZ.** Locate the Bright Angel Shale (Cba) highlighted by the Problem 5 placemark. What kind of slope does it form?
   a. steep cliff
   b. gentle slope
6. **Differential Weathering - Grand Canyon NP, AZ.** Locate the Coconino Sandstone (Pc) highlighted by the Problem 6 placemark. What kind of slope does it form?
   a. steep cliff
   b. gentle slope
7. **Differential Weathering - Grand Canyon NP, AZ.** What general statement can you now make about the slopes of the Grand Canyon based on the different <u>sedimentary rock types</u>?
   a. all sedimentary rocks weather to form uniform slopes
   b. there is no consistent slope pattern related to type of sedimentary rock
   c. "strong" rocks (shales) form cliffs, "weak" rocks (limestones/sandstones) form slopes
   d. "strong" rocks (limestones/sandstones) form cliffs, "weak" rocks (shales) form slopes

As we discovered in the Grand Canyon, one of the most easily observable sedimentary structures is **bedding** or **layering**. However, not all regions have experienced the spectacular erosion that the Grand Canyon has. In mountainous areas, plate tectonic convergence has exposed various sedimentary layers by tilting, folding, and/or faulting them. Resistant rocks still erode to form cliffs, and less resistant rocks still form low slopes or valleys. Folding and faulting can repeat the same sedimentary layers multiple times, creating a series of repeating ridges and valleys.

8. **Differential Weathering - Lewis Range, MT.** Which of the following is true regarding the resistance to erosion of the layers specified by the Problem 8 placemarks?
   a. Problem 8a is less resistant and Problem 8b is more resistant
   b. Problem 8a is more resistant and Problem 8b is less resistant
   c. Problem 8a and Problem 8b have similar resistances to weathering and erosion
9. **Physical Weathering - Lewis Range, MT.** In temperate climates, water can infiltrate cracks in the sedimentary rocks (and other types as well), and with repeated freeze-thaw cycles, can break the rocks apart by the **physical weathering** process of **frost/ice wedging**. The broken rocks often fall down the steep slope to form a pile of rubble at the base of the cliff. What is this pile of broken rocks called?
   a. sediment
   b. moraine
   c. talus
   d. till

Sedimentary rocks form in a variety of **terrestrial** (non-marine) and **marine/coastal** depositional environments. The **texture** and **composition** of the resulting sedimentary rocks reflects the processes operative in that particular environment and the type of materials available.

10. **Modern Depositional Environments - Death Valley NP, CA.** What depositional environment is highlighted by the Problem 10 placemark?
    a. sand dunes *(forming sandstone)*
    b. alluvial fan *(forming arkose & conglomerate)*
    c. playa lake *(forming evaporite rocks)*
    d. mountain stream *(forming large boulder/cobble conglomerates)*
11. **Modern Depositional Environments - Death Valley NP, CA.** What depositional environment is highlighted by the Problem 11 placemark?
    a. sand dunes *(forming sandstone)*
    b. alluvial fan *(forming arkose & conglomerate)*
    c. playa lake *(forming evaporite rocks)*
    d. mountain stream *(forming large boulder/cobble conglomerates)*

12. **Modern Depositional Environments - Niger Delta, Nigeria.** Fly to the Niger Delta in Nigeria, and turn on the *Niger Delta* folder to activate a polygon that provides the approximate present-day outline of the delta with features created by running water individually labeled. What type of sedimentary rock will likely form at each of the placemarks in their present-day setting?
   a. 12a: sandstone/conglomerate and 12b: shale/siltstone/coal
   b. 12a: shale/siltstone/coal and 12b: sandstone/conglomerate
13. **Modern Depositional Environments - Niger Delta, Nigeria.** If you owned the company at the Problem 13 placemark, which of the following concerns would you have?
   a. the meander would continue migrating <u>toward</u> my buildings
   b. the meander would continue migrating <u>away from</u> my buildings
14. **Modern Depositional Environments - Grand Canyon NP, AZ.** Here, the Colorado River narrows at the mouth of a side canyon, and there are some rapids. What causes this?
   a. a **delta** containing sand and gravel from a side canyon
   b. a **point bar** of sand and gravel at the inside bend of a meandering river
   c. fine silts from a river **floodplain**
15. **Ancient Depositional Environments - Grand Canyon NP, AZ.** At this location, note the following sequence of Cambrian rock units: Tapeats Sandstone (near shore; lowest rock unit), Bright Angel Shale (intermediate distance from shore; middle rock unit), and Muav Limestone (distal from shore sediments, but still shallow water; top rock unit). The vertical stacking of these units represents either a **regression** (fall) or **transgression** (rise) of sea level. Which is occurring in this area? *Hint: Draw an initial shoreline showing the sandstone (near), shale (intermediate), and limestone (distal/far) deposits, and then draw a second higher shoreline (transgression) with the same distribution of rock units. Compare the vertical stacking of transgressive rock layers with the Cambrian rock units in the Grand Canyon to determine if the sequence there is transgressive or regressive (add a third higher shoreline if necessary).*
   a. regression *(fall in sea level)*
   b. transgression *(rise in sea level)*
16. **Ancient Depositional Environments - Checkerboard Mesa, Zion NP, UT.** Checkerboard Mesa beautifully exposes a large outcrop of mostly stark white rock that was deposited in an aeolian (desert) environment during the Jurassic Period. What type of rock is present?
   a. limestone
   b. shale
   c. sandstone
   d. breccia
17. **Ancient Depositional Environments - Checkerboard Mesa, Zion NP, UT.** Checkerboard Mesa gets its name from the intersection of sub-vertical fractures/joints with more gently tilted layers. Knowing that these are lithified sand dunes, what sedimentary structure forms these tilted layers *(trace an individual tilted layer upward and note that it becomes truncated)*?
   a. cross-beds
   b. ripple marks
   c. mudcracks
   d. graded beds

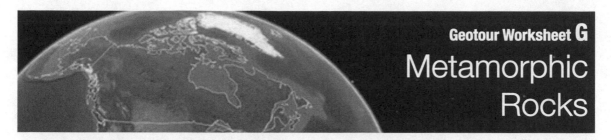

Metamorphic Rocks

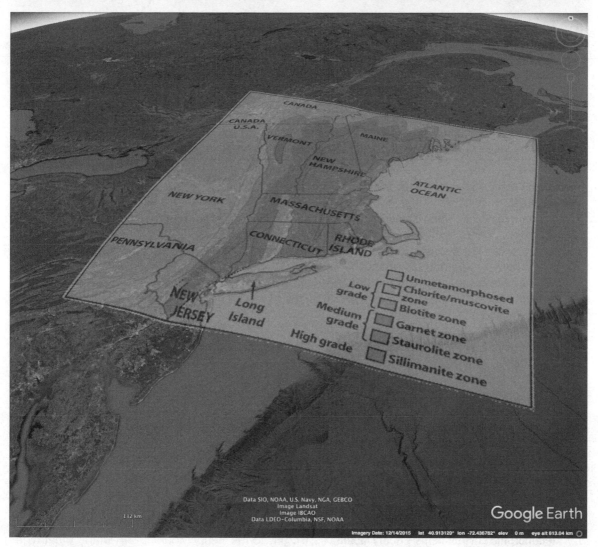

The central New England area (42.676°N, 071.995°W) exposes the central metamorphic core of the Paleozoic-age Appalachian Mountains. In the image above, a georeferenced map shows the metamorphic zones based on mapping the occurrence of various metamorphic index minerals. As you travel west from Massachusetts and Connecticut, you quickly cross into areas of non-metamorphic rock in New York and Pennsylvania.

> *Circle the letter next to the best answer for each question.*
>
> **Worksheet Resources:**
> - **Google Earth** - Open the **2. Exploring Geology Using Geotours > G. Metamorphic Rocks** folder.
>   1. **Problem Materials** - **Check and double-click items associated with each problem** to travel to the appropriate location with the prescribed perspective/zoom.
>   2. **Metamorphic Rocks Geotours Library** - **Explore additional Geotours** in this folder to help answer problems.
> - **Geoscience** - **Consult a textbook** and/or **Internet resources** to help answer some problems.

Turn on the *New England Metamorphic Zones* overlay, and double-click it to fly to the area.

1. **Metamorphic Zones - New England, USA.** If you traveled north from the Problem 1 placemark, would you go into higher or lower **metamorphic grade** rocks?
   a. lower metamorphic grade
   b. higher metamorphic grade

2. **Metamorphic Zones - New England, USA.** Based on your answer to (a), would the **metamorphic core** *(the zone of highest metamorphism)* of this region be near placemark Problem 2a or 2b?
   a. Problem 2a placemark
   b. Problem 2b placemark

3. **Metamorphic Zones - New England, USA.** For the schist pictured in the Problem 3 placemark, which zone did it likely occur in *(identify the red minerals)*?
   a. unmetamorphosed
   b. chlorite/muscovite zone
   c. biotite zone
   d. garnet zone

4. **Metamorphic Environments - New England, USA.** Considering the area covered by the *New England Metamorphic Zones* overlay, what type of metamorphism likely produced these metamorphic rocks?
   a. shock metamorphism *(e.g., impact of a meteorite)*
   b. contact metamorphism *(e.g., intrusion of igneous sills & dikes)*
   c. regional metamorphism *(e.g., burial and/or large-scale convergent tectonics)*

5. **Metamorphic Environments - Henry Mountains, UT.** Fly to the Henry Mountains in Utah. What type of metamorphism likely produced the thin, white bleaching of the reddish-tan sandstone in this area? *Hint: You may want to fly around this area to look at the features present.*
   a. shock metamorphism *(e.g., impact of a meteorite)*
   b. contact metamorphism *(e.g., intrusion of igneous sills & dikes)*
   c. regional metamorphism *(e.g., burial and/or large-scale convergent tectonics)*

6. **Metamorphic Environments - Van Hise Rock, Rock Springs, WI.** Fly to the Rock Springs, WI area to view this outcrop using Street View. After the sediment composing these rocks was initially deposited and lithified, these Precambrian rocks experienced greenschist grade metamorphism as the sedimentary rocks were covered with additional rock layers and then subsequently folded and sheared past each other. Which of the choices below best describes the type of metamorphism that these rocks experienced?
    a. shock metamorphism *(e.g., impact of a meteorite)*
    b. contact metamorphism *(e.g., intrusion of igneous sills & dikes)*
    c. regional metamorphism *(e.g., burial and/or large-scale convergent tectonics)*

7. **Protoliths - Van Hise Rock, Rock Springs, WI.** The **protolith** (parent rock) of the pink/maroon rock at Van Hise Rock was a <u>pure</u> quartz sandstone (right side with orange field book), whereas the darker layer was originally a lens of clay-rich shale (middle). What metamorphic rocks did these protoliths (sandstone, shale) become? *You may need to refer to your textbook.*
    a. quartzite, marble
    b. quartzite, phyllite
    c. slate, marble
    d. marble, gneiss

Turn on the *Global Map of Cratonic Shields & Mt Belts* overlay to view the Precambrian **cratonic shields** (maroon) and the younger **mountain belts** (green). Also, in the *Layers* panel, turn on *Gallery > Earthquakes* and *Gallery > Volcanoes*.

8. **Cratonic Shields & Mountain Belts - Andes Mountains, South America.** Fly in to the Andes mountains on the South American continent until you see the earthquake and volcano placemarks. Using these as a proxy for tectonic activity, which of the statements below best describes your observations *(these generally hold for all Precambrian shields and younger mountain belts)*?
    a. Precambrian shields are less tectonically active than younger mountain belts
    b. Precambrian shields are more tectonically active than younger mountain belts
    c. Precambrian shields and younger mountain belts are similarly tectonically active

9. **Cratonic Shields & Mountain Belts - Canadian Shield, Hudson Bay, Canada.** Turn off the *Global Map of Cratonic Shields & Mt Belts* overlay, and fly to the Canadian Shield near Hudson Bay. Here, **metamorphic foliations** have differentially eroded to show two distinct foliation trends of different ages (one truncating the other). What evidence do you see in the foliations to suggest that these rocks were once deeply buried and experienced tectonic deformation?
    a. certain rock layers have developed basins that now contain water
    b. there is a distinct cleavage in the rocks when viewed closely
    c. the foliations are deformed into contorted folds like we see in many gneisses
    d. these rocks show no evidence of tectonic deformation

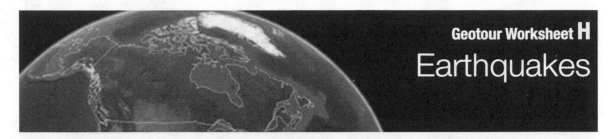

Earthquakes

The image above shows occurrences of recent seismicity in the western United States. While there are numerous earthquakes on faults throughout the Rocky Mountain and Basin & Range areas, most of the earthquakes are focused along the San Andreas Fault (35.169°N, 119.717°W), the transform boundary separating the North American Plate from the Pacific Plate. Other seismic events are located throughout the Pacific Plate with concentrations along the divergent spreading ridge along the western margin of the Juan de Fuca Plate, a small remanent of the larger Farallon Plate that is subducting beneath the NW coast of the United States.

Name:_____

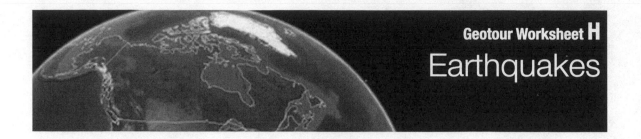

Geotour Worksheet H
Earthquakes

Circle the letter next to the best answer for each question.

Worksheet Resources:
- **Google Earth** - Open the **2. Exploring Geology Using Geotours > H. Earthquakes** folder.
  1. **Problem Materials** - **Check and double-click items associated with each problem** to travel to the appropriate location with the prescribed perspective/zoom.
  2. Earthquakes Geotours Library - **Explore additional Geotours** in this folder to help answer problems.
- **Geoscience** - **Consult a textbook** and/or **Internet resources** to help answer some problems.

Note: Images for the Locating Earthquakes & Determining Richter Magnitudes exercises used with permission from the Virtual Courseware Project (ScienceCourseware.org) at California State University, Los Angeles, CA.

Earthquakes generate vibrations through the earth, which are recorded by seismograph stations around the world in the form of *xy* graphs called **seismograms** (see labeled simulated seismogram to the right). To estimate the location of the earthquake's **epicenter**, one must use the seismograms from at least three seismographs to determine the time interval between the arrival of the S and P waves (called the **S-P interval** or **S-P lag time**; 36 seconds in the simulated seismogram).

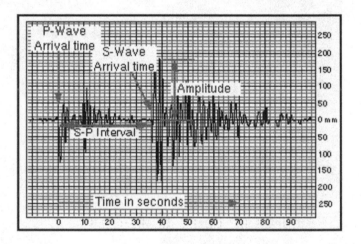

1. **Locating Earthquakes - Tokyo, Japan.** Using the simulated seismogram in the Problem 1 placemark from the Tokyo, Japan seismograph station, determine the **S-P interval/lag time** (sec) for the earthquake waves to travel from the epicenter to Tokyo.
   - a. ~62 sec
   - b. ~44 sec
   - c. ~28 sec
   - d. ~15 sec
2. **Locating Earthquakes - Akita, Japan.** Using the simulated seismogram in the Problem 2 placemark from the Akita, Japan seismograph station, determine the **S-P interval/lag time** (sec) for the earthquake waves to travel from the epicenter to Akita.
   - a. ~71 sec
   - b. ~37 sec
   - c. ~19 sec
   - d. ~5 sec

3. **Locating Earthquakes - Busan, South Korea.** Using the simulated seismogram in the Problem 3 placemark from the Busan, South Korea seismograph station, determine the **S-P interval/lag time** (sec) for the earthquake waves to travel from the epicenter to Busan.
    a. ~74 sec
    b. ~56 sec
    c. ~27 sec
    d. ~39 sec

4. **Locating Earthquakes - Tokyo, Akita, and Busan, SE Asia.** Using the S-P interval/lag time curves in the Problem 4 placemark and the answers to Problems 1, 2, & 3, determine the distance (km) from each of the seismograph stations to the epicenter.
    a. Tokyo (~198 km), Akita (~300 km), & Busan (~240 km)
    b. Tokyo (~290 km), Akita (~190 km), & Busan (~260 km)
    c. Tokyo (~600 km), Akita (~690 km), & Busan (~720 km)
    d. Tokyo (~430 km), Akita (~695 km), & Busan (~545 km)

5. **Locating Earthquakes - Tokyo, Akita, and Busan, SE Asia.** Using the *Circle* tab on the *Ruler* tool, draw a circle for each of the seismograph stations with radii corresponding to your answer for Problem 4. <u>Save each circle after drawing them</u>. Which placemark correctly identifies where the epicenter is located?
    a. Problem 5a
    b. Problem 5b
    c. Problem 5c
    d. Problem 5d

To estimate the **Richter magnitude** for an earthquake, one must know the **distance** (km; e.g., answer for Problem 4) of a seismograph station from the earthquake epicenter and the **maximum amplitude** (mm; ~180 mm in the previous simulated seismogram above) of the S wave on that station's seismogram. Using a **Richter nomogram** (right; essentially a graph with three axes scaled based on Richter's equations to account for distance from the earthquake epicenter), the magnitude can be estimated by drawing a line from the **distance axis** to the S-wave **amplitude axis** using the values for that seismograph station. Where the line crosses the **magnitude axis** is the Richter magnitude for the earthquake. Again, it is best to use seismograms from at least three seismographs to converge on a single best answer.

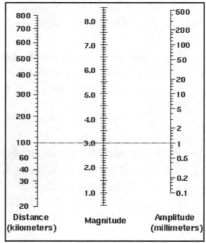

6. **Determining Richter Magnitude - Tokyo, Japan.** Using the simulated seismogram in the Problem 6 placemark from the Tokyo, Japan seismograph station, determine the **maximum S-wave amplitude** (mm) for the seismogram.
    a. ~210 mm
    b. ~100 mm
    c. ~130 mm
    d. ~170 mm

7. **Determining Richter Magnitude - Akita, Japan.** Using the simulated seismogram in the Problem 7 placemark from the Akita, Japan seismograph station, determine the **maximum S-wave amplitude** (mm) for the seismogram.
    a. ~30 mm
    b. ~21 mm
    c. ~71 mm
    d. ~5 mm

8. **Determining Richter Magnitude - Busan, South Korea.** Using the simulated seismogram in the Problem 8 placemark from the Busan, South Korea seismograph station, determine the **maximum S-wave amplitude** (mm) for the seismogram.
   a. ~65 mm
   b. ~90 mm
   c. ~56 mm
   d. ~18 mm

9. **Determining Richter Magnitude - Tokyo, Akita, and Busan, SE Asia.** Using the *Problem 9 - Richter nomogram* overlay *(not the one in the overlay balloon)* and the *Line* tab of the *Ruler* tool, draw a line for each seismograph station using the distance from each seismograph to the earthquake epicenter (from Problem 4 in km; left axis) and the S-wave amplitude measured from its seismogram (in millimeters; right axis). Save each line after drawing them. Estimate the Richter magnitude based on where the lines for the three seismographs cross the center axis of the Richter nomogram (they may not intersect perfectly). What is the Richter magnitude for this earthquake?
   a. ~5.8-6.0
   b. ~6.1-6.3
   c. ~4.5-4.7
   d. ~6.7-6.9

One of the most active (and best studied) earthquake-prone areas is the San Andreas transform fault that forms a **plate boundary** between the North American Plate and the Pacific Plate. Turn on the *San Andreas Major Ruptures* folder to see rupture extents for three major earthquakes that have affected the San Andreas: Fort Tejon (1857, Mag. 7.9), San Francisco (1906, Mag. 7.8), and Loma Prieta (1989, Mag. 6.9). Also, in the *Layers* panel, turn on *Gallery > Earthquakes*.

10. **Plate Boundary Earthquakes - San Andreas Fault, Wallace Creek, CA.** What is the sense of offset along the San Andreas Fault? *(i.e., Walk along either segment of Wallace Creek toward the fault. At the fault, determine the direction you turn to find the offset segment of the stream across the fault.)*
    a. right lateral *(rocks on the west side of the San Andreas Fault move to the NW)*
    b. left lateral *(rocks on the west side of the San Andreas Fault move to the SE)*

11. **Plate Boundary Earthquakes - San Andreas Fault, Wallace Creek, CA.** Use the *Ruler* tool to determine the distance (m) that Wallace Creek has been offset by recent motion along the San Andreas Fault (measure between the Problem 11 placemarks).
    a. ~75 m
    b. ~100 m
    c. ~303 m
    d. ~160 m

12. **Plate Boundary Earthquakes - San Andreas Fault, Wallace Creek, CA.** Assuming that slip accumulates along the fault at about 6 cm/yr, how many years might it have taken to accumulate the slip that was released to produce the observed offset of the stream?
    a. ~2667 yrs
    b. ~5033 yrs
    c. ~1010 yrs
    d. ~10,100 yrs

13. **Plate Boundary Earthquakes - San Andreas Fault, Palmdale, CA.** Fly to the Palmdale section of the San Andreas Fault and turn on the *Offset Stream* path. Use the *Ruler* tool to determine the distance (m) that this creek has been offset by recent motion along the San Andreas Fault (measure between the Problem 13 placemarks). Assuming that slip accumulates along the fault at about 6 cm/yr, how many years might it have taken to accumulate the slip that was released to produce the observed offset of the stream?
    a. 20,000-25,000 yrs
    b. 8000-10,000 yrs
    c. 4000-5000 yrs
    d. 800-1000 yrs

14. **Plate Boundary Earthquakes - San Andreas Fault, Palmdale, CA & Wallace Creek, CA.** Both of these areas were affected by the Fort Tejon earthquake that slipped a maximum of 9 m (30 ft) in a single event in 1857 (make sure the *San Andreas Major Ruptures* folder is turned on). Given your answers to Problems 11-13, which of the following is <u>incorrect</u>?
    a. the 1857 Fort Tejon earthquake cannot be the sole cause for the offsets of either stream
    b. given the times necessary to accumulate slip for each location, one possible interpretation is that Wallace Creek is much younger than the creek at Palmdale and has not recorded as much offset
    c. these offset streams record the total amount of slip occurring along the San Andreas
    d. adjacent faults might accommodate some of the slip between the two areas

15. **Plate Boundary Earthquakes - San Andreas Fault, CA.** Keep the *San Andreas Major Ruptures* folder turned on to show the extents of segments of the San Andreas Fault that have slipped during major seismic events. Using **seismic gap** analysis (where segments of a fault that have not recently slipped form "gaps" of accumulating strain transferred from adjacent slipped fault segments that highlight areas of greater potential for a "seismic" event), which placemark highlights a segment of the San Andreas Fault that has a high risk for an earthquake?
    a. Problem 15a
    b. Problem 15b
    c. Problem 15c
    d. Problem 15d

Turn on the *NMSZ Structural Features* overlay and double-click it to fly to the Missouri, Illinois, Kentucky, Arkansas, and Tennessee area (if you would like to see modern seismicity, turn on *Gallery > Earthquakes* in the *Layers* panel as well). This area is well within the North American Plate, yet it has a surprising number of earthquakes (the densest region of earthquakes is referred to as the **New Madrid Seismic Zone**, which helps delineate the **Reelfoot Rift** (black lines), a continental rift that failed to continue to extend to form a plate boundary...such failed rifts are called **aulacogens**). The NMSZ was the site of three large (~7.0 to 7.5) earthquakes (and one major aftershock) from December 1811 through February 1812.

16. **Intraplate Earthquakes - New Madrid, MO.** With the *NMSZ Structural Features* overlay on, you should note three distinct segments of the NMSZ: **New Madrid North Zone** (~Paducah, KY to New Madrid, MO), **Reelfoot Central Zone E & W** (~Bernie, MO/New Madrid, MO to Dyersburg, TN), and **Blytheville South Zone** (~Caruthersville, MO to Marked Tree, AR). What is the dominant trend/pattern of earthquakes in the <u>north</u> (yellow lines) and <u>south (green lines defining the Blytheville Arch) zones</u>?
    a. E-W
    b. N-S
    c. NE-SW
    d. NW-SE

17. **Intraplate Earthquakes - New Madrid, MO.** What is the dominant trend/pattern of earthquakes in the <u>central zone</u>?
    a. E-W
    b. N-S
    c. NE-SW
    d. NW-SE

18. **Intraplate Earthquakes - New Madrid, MO.** Now turn on the *1811-1812 Epicenters* folder to see the estimated locations for the major earthquakes that occurred during the winter of 1811-1812. While the faults responsible for these major earthquakes likely originated as normal faults during formation of the Reelfoot Rift (black lines), they now accrue slip in response to the regional contractional stresses shown by the large red arrows. What is the current sense of slip in the <u>north</u> and <u>south zones</u>? *Note: The 12/16/1811 & 1/23/1812 earthquakes slipped in this fashion. In addition, the black lines that define the Reelfoot Rift and the red lines that define the Ridgely fault zone are sub-parallel to the north and south zones.*
    a. NW-SE extension (pulling apart ~perpendicular to the yellow and green lines)
    b. NW-SE contraction (shortening ~perpendicular to the yellow and green lines)
    c. NE-SW transpression (a combination of <u>trans</u>form-like strike-slip and com<u>pression</u> at a low angle to the yellow and green lines)

19. **Intraplate Earthquakes - New Madrid, MO.** What is the sense of slip on the <u>red fault line marked as the Reelfoot Fault</u> in the <u>central zone</u>? *Note: The 2/07/1812 earthquake slipped in this fashion.*
    a. NW-SE extension (pulling apart ~perpendicular to the red Reelfoot Fault)
    b. NW-SE contraction (shortening ~perpendicular to the red Reelfoot Fault)
    c. NE-SW transpression (a combination of <u>trans</u>form-like strike-slip and com<u>pression</u> at a low angle to the red Reelfoot Fault)

20. **Intraplate Earthquakes - New Madrid, MO.** Movement on the Reelfoot Fault (red line) during the 2/07/1812 earthquake uplifted one side of the fault by as much as 3 m (9 ft), causing temporary waterfalls on the Mississippi River to alter the normal direction of flow on a segment of the river and damming up a stream in the Mississippi River floodplain to form Reelfoot Lake (Problem 20 placemark). Which side of the Reelfoot Fault was <u>uplifted</u>?
    a. ENE
    b. WSW

21. **Intraplate Earthquakes - New Madrid, MO.** Turn off the *1811-1812 Epicenters* folder and the *NMSZ Structural Features* overlay. The intense shaking by the 1811-1812 earthquakes produced a large area of circular and oval deposits of sand (especially in areas covered by deposits left behind as the Mississippi River changed its course over time; toggle the *NMSZ Liquefaction Features* folder on and off to see the affected areas). The Problem 21 placemark takes you to an exposure of such a deposit. What is this feature called?
    a. sand scarp
    b. sag pond
    c. tsnumai deposit
    d. sand blow

Check and double-click the *North Anatolian Fault, Turkey* folder to fly to northern Turkey. The segments within the folder show the traces of a series of faults that compose the North Anatolian Fault system. The segments are labeled and colored to show a relatively brief historical record of earthquake activity along the North Anatolian Fault (based on approximate regions that were active during each major seismic event...gray faults have not been active during these major events).

22. **Earthquake Prediction - North Anatolian Fault, Turkey.** What is the dominant pattern of major earthquakes along the North Anatolian Fault from <u>1939 to 1999</u>?
    a. generally become progressively younger from west (oldest) to east (youngest)
    b. generally become progressively younger from east (oldest) to west (youngest)
    c. there is no apparent pattern

23. **Earthquake Prediction - North Anatolian Fault, Turkey.** What might cause this pattern?
    a. slip from the rupture of one earthquake increased stress on adjacent fault segments
    b. slip from the rupture of one earthquake decreased stress on adjacent fault segments
    c. there is no apparent pattern, so the earthquakes occurred randomly along the fault

24. **Earthquake Prediction - North Anatolian Fault, Turkey.** Assuming you are a geologist/geophysicist consulting with the Turkish government, where would you predict would be the likeliest area to next experience a major earthquake in association with this fault system?
    a. Istanbul area (Problem 24a placemark)
    b. Osmancik area (Problem 24b placemark)
    c. Alaca area (Problem 24c placemark)
    d. Duzce area (Problem 24d placemark)

Check and double-click the *Regional Plate Tectonics* map overlay and the *Epicenter* placemark to fly to the Sumatra region that was devastated by a tsunami in late 2004. Also, in the *Layers* panel, turn on *Gallery > Earthquakes* and *Gallery > Volcanoes*.

25. **Tsunami Devastation - Sumatra.** Which tectonic plate was subducted beneath which overriding tectonic plate at this locality to produce this massive earthquake? *Note that the subduction zone is delineated by the line with triangles and that the volcanoes will be on the overriding plate.*
    a. Indian Plate is being subducted beneath the Burma Plate
    b. Burma Plate is being subducted beneath the Indian Plate
    c. Sunda Plate is being subducted beneath the Burma Plate
    d. Burma Plate is being subducted beneath the Sunda Plate

26. **Tsunami Devastation - Sumatra.** Check and double-click the *2004 Indian Ocean Tsunami Propagation Animation* placemark, and click on it to open its balloon and watch how the tsunami propagated from the epicenter across the Indian Ocean. Close this placemark, and then check and double-click the *2004 Indian Ocean Tsunami Travel Times* overlay. How long did it take the tsunami to first arrive at the Problem 26 placemark on Sri Lanka? *Note that each line represents 1 hour.*
    a. 1 hr
    b. 1.5 hr
    c. 2 hr
    d. 5 hr

27. **Tsunami Devastation - Sumatra.** Use the *Ruler* tool to determine the distance (km) from the earthquake epicenter to Sri Lanka (measure between the Problem 27 placemarks). Use your answer from Problem 26 to determine the average velocity at which the tsunami traveled across the Indian Ocean (km/hr).
    a. 150-275 km/hr
    b. 750-900 km/hr
    c. 350-550 km/hr
    d. 1250-1400 km/hr

28. **Tsunami Devastation - Sumatra.** The Problem 28 placemarks show three different areas on the northern tip of Sumatra before and after the devastating tsunami. Click each placemark and fly over the region to see the extent of the tsunami damage. Based on these images, where is the safest place to be along tsunami-prone coastal areas?
    a. buildings (water will go around buildings)
    b. high ground (preferably inland from the coast)
    c. flat coastal areas (water can drain away better)
    d. beachfront homes on stilts (water can go beneath house)

The image above features a slightly off-axis view of the NW-plunging nose of the Sheep Mountain anticline (44.642°N, 108.186°W) in the Bighorn Basin of Wyoming (left side of image). You can recognize that this basement-involved fold is plunging by continuously tracing rock layers (e.g., the red-orange Triassic Chugwater Formation) from one fold limb across the fold axis to the other fold limb without interruption. Flatirons (triangular landforms that resemble laundry irons) characterize each fold limb, with the tip of the flatiron pointing in the direction opposite of the dip of the rock layers. Flatirons on the NE flank of Sheep Mountain indicate a relatively steep dip to the NE, whereas flatirons on the far right side of the image show a relatively gentle dip to the SW. Together, these dips suggest that the Sheep Mountain anticline is part of an anticline-syncline pair of folds in this area.

Geotour Worksheet I
Geologic Structures

Circle the letter next to the best answer for each question.

Worksheet Resources:
- **Google Earth** - Open the **2. Exploring Geology Using Geotours > I. Geologic Structures** folder.
  1. **Problem Materials** - **Check and double-click items associated with each problem** to travel to the appropriate location with the prescribed perspective/zoom.
  2. Geologic Structures Geotours Library - **Explore additional Geotours** in this folder to help answer problems.
- **Geoscience** - **Consult a textbook** and/or **Internet resources** to help answer some problems.

Tilted and/or folded sedimentary rocks typically erode to form asymmetric ridges called **flatirons**— relatively planar surfaces usually representing **bedding planes** whose tip points in the direction opposite of the tilt *(geoscientists thought that this geometry resembled the flat bottom of a laundry iron, hence the name)*. Geoscientists describe the orientation of such structures by measuring the **strike** and **dip** of rock layers that compose the structure. **Strike** is the angle between an imaginary horizontal line on the structure *(often visualized as the intersection of the structure's slope and the flat surface of a body of water)* and the direction of true north, whereas **dip** is the angle of a structure's slope measured in a vertical plane perpendicular to the strike *(the **dip direction** is often visualized as the direction in which water poured on the structure's slope would flow)*.

1. **Flatirons - Wind River Mountains, WY.** The Problem 1 polygon highlights a single flatiron in a tilted region composed of numerous flatirons (some small with narrow, pointed tips and some large with broad, rounded tips) on the northeastern flank of the Wind River Mountains uplift in Wyoming. Rotate the view so that you are looking directly down on the flatiron with north at the top. Which direction are the rock layers dipping? *To make it easier to see, you might consider temporarily increasing the vertical exaggeration to 2 in Google Earth.*
    - a. NE
    - b. NW
    - c. SE
    - d. SW

2. **Strike - Wind River Mountains, WY.** Check and double-click the Problem 2 placemark. To approximate strike, use the *Line* feature in the *Ruler* tool (make sure that the *Ruler* tool and the *Status Bar* are using meters) to determine the elevation at the point of the Problem 2 placemark icon by hovering the cursor over it and looking at the *Status Bar* (~1835 m). Now draw a line from the Problem 2 placemark using the *Ruler* tool to the right (NW) such that you click the second point of the line at a location at the exact same elevation (~1835 m). Use the *heading of this horizontal line in the Ruler tool dialog box* to determine the orientation of strike (000°-360°) for these rock layers. *Note: Directions can be visualized as a 360° circle with N = 000° and 360°, E = 090°, S = 180°, and W = 270°. This approximation works because we're assuming that the flat part of the flatiron approximates a bedding plane surface.*
    - a. 230°-240°
    - b. 040°-050°
    - c. 310°-320°
    - d. 125°-135°

3. **Dip Direction - Wind River Mountains, WY.** Leave the 1835 m line on the screen and turn on the Problem 3 placemark. To determine dip direction, draw a perpendicular line <u>from the 1835 m line to the Problem 3 placemark point</u> (use the *Ruler* tool; the strike line will disappear and you will now see only the dip line). The orientation of the dip line is the dip direction (the direction that water would flow down the flatiron). What is the dip direction (000°-360°)?
   a. 040°-050°
   b. 170°-180°
   c. 210°-220°
   d. 310°-320°

4. **Dip - Wind River Mountains, WY.** Note the map length of the dip line in meters from the *Ruler* tool dialog box and hover the cursor over the point of the Problem 4 placemark to determine its elevation. What is the approximate dip angle (00°-90°)? To determine the dip angle, use the following formula *(note that your elevation and length units must all be meters and that your calculator must be in degrees and not radians)*:

   dip angle = tan⁻¹ [(strike line elevation - Problem 4 placemark elevation) ÷ map length of the dip line].
   a. 05°-12°
   b. 19°-26°
   c. 31°-37°
   d. 43°-48°

<u>Just for Fun</u>...Determining the strike and dip of a flatiron is harder to explain in words than it is to actually do in Google Earth! Find some other nearby flatirons (look for the most planar ones) and try to draw a horizontal strike line. How does it compare to your answer? Do the same for dip and see how close you come. Think about reasons why you might have differences.

Check and double-click the *Arches NP Geologic Map-North* overlay and make it semi-transparent. A road (dashed red/brown line in the yellow Qea unit on the geologic map) approximates the axis of a salt-filled anticline that has collapsed to form Salt Valley. Along the flanks of this anticlinal valley, some of the rock units exhibit a prominent linear pattern that reflects systematic fracture sets that have developed in association with the Salt Valley anticline and subsequent collapse.

5. **Fractures - Arches National Park (NP), UT.** The Problem 5 placemark points to the youngest rock unit that can be easily observed to contain these fracture sets. Using the geologic map overlay, which unit is this? *Note: The oldest rock unit in the map key is the Paradox Formation and the youngest is the Modern Alluvial Deposits (older rock units are listed on the bottom and younger ones are on top). Make sure that your map is semi-transparent or simply toggle it on/off.*
   a. Jcd
   b. Jes
   c. Jctm
   d. Jmt

6. **Fractures - Arches NP, UT.** What happens to the fracture sets beneath the younger rock units (Jmt & Jms) at the Problem 6 placemark?
   a. they do not exist beneath the Problem 6 placemark anywhere in the subsurface
   b. they continue on beneath the Problem 6 placemark in the white unit (Jctm) because the younger Jmt/Jms units just cover them

<u>Just for Fun</u>...Think about your answer for Problem 6...does this help date the age of jointing or does it just reflect how joints develop differently in different rock types?

7. **Fractures - Arches NP, UT.** Check and double-click the placemark for Problem 7, which flies you to an area on the NE side of Salt Valley that shows two prominent joint sets. Joint sets may intersect at a variety of angles, but most commonly are classified as being at right angles (orthogonal) or not at right angles. Which kind are these?
   a. right angles (orthogonal)
   b. not at right angles

8. **Fractures - Arches NP, UT.** Arches National Park is known for its spectacular arches that develop in the jointed areas of the park. The Problem 8 placemark flies you to Landscape Arch, the arch with the largest span in Arches National Park. If the stresses that stretched the rock to form the joints were oriented <u>perpendicular</u> to the joints (and the rock fins that contain the arches), in what direction were the stresses oriented?
   a. N-S
   b. E-W
   c. NW-SE
   d. NE-SW

Faults are typically classified in one of three main categories: **normal**, **reverse**, and **strike-slip**. The following questions will take you to locales that provide examples of each type of these faults.

9. **Normal Faults - Canyonlands NP, UT.** Normal faults in the Grabens area in Utah's Canyonlands National Park–Needles District accommodate regional extension above the underlying Paradox Formation (a thick interval of evaporite deposits, including rock salt and gypsum). This area is referred to as "the Grabens" because faulting forms fault-bounded blocks that have been down-dropped (**grabens**) with intervening high blocks (**horsts**). Which of the following is <u>correct</u>?
   a. Problem 9a placemark is a graben
   b. Problem 9b placemark is a graben
   c. Problem 9c placemark is a graben
   d. Problem 9d placemark is a graben

10. **Normal Faults - Squaw Point, OR.** The Problem 10 placemark lies on a normal **fault scarp**, an offset of the land surface by slip along a fault. The **hanging wall** lies above the fault surface, whereas the **footwall** lies below the fault surface (i.e., if you were to walk on the fault, your feet would be on the footwall). For normal faults, the hanging wall moves down relative to the footwall to accommodate extension. Which side of the normal fault is the down-dropped hanging wall?
    a. Problem 10a placemark
    b. Problem 10b placemark

11. **Normal Faults - Short Lake, OR.** The Problem 11 placemark highlights another area east of the Problem 10 placemark that has experienced extension along a series of **synthetic normal faults** (faults that all dip in the same direction), forming a series of **half grabens**. If the extensional stresses that are stretching the rocks are approximately perpendicular to the fault traces, which direction are the stresses oriented? *Note: It may be useful to orient the view to look directly down from above with north to the top.*
    a. N-S
    b. E-W
    c. NW-SE
    d. NE-SW

12. **Reverse (Thrust) Faults - Canadian Rockies, Alberta, Canada.** The Problem 12 photo shows a nearly cross-sectional view of a low-angle reverse fault (the McConnell thrust fault) placing older Paleozoic rocks (gray) on top of forested Mesozoic rocks. The gray rocks are interpreted to have been uplifted over 5 km and transported more than 40 km from their original site of deposition. Which direction were the rocks transported? *You might also turn on the Kananaskis Country Geologic Map, Canada map overlay and adjust the photo's transparency (triangles are on the hanging-wall side of the fault).*
   a. to the right (east)
   b. to the left (west)

13. **Reverse (Thrust) Faults - Canadian Rockies, Alberta, Canada.** Check and double-click the placemarks for Problem 13, and turn on the *Kananaskis Country Geologic Map, Canada* overlay. Reverse faults, and particularly thrust faults, duplicate rock units by stacking the same units on top of each other over and over. How many <u>major</u> repetitions of the Devonian Palliser Formation (DPa, cyan colored) are seen in the rocks north of the Bow River? *Count the repetitions marked by the placemarks.*
   a. zero
   b. one
   c. three
   d. nine

14. **Reverse (Thrust) Faults - Canadian Rockies, Alberta, Canada.** The Problem 14 photo is almost a cross-sectional view that shows a zoomable photo of a reverse fault near Barrier Lake, Canada (turn on the *Kananaskis Country Geologic Map, Canada* overlay and adjust the photo's transparency). Which direction were the rocks above the fault displaced?
   a. to the right (east)
   b. to the left (west)

15. **Strike-Slip Faults - Carrizo Plain, CA.** The Problem 15 placemarks highlight a stream near the Carrizo Plain in California that has been offset along the San Andreas fault zone (the San Andreas Fault is a strike-slip fault that forms a transform plate boundary between the North American and Pacific plates). Given the offset stream in the view, what is the sense of offset along this fault?
   a. left-lateral
   b. right-lateral

16. **Strike-Slip Faults - Altyn Tagh Fault, China.** The Problem 16 placemark highlights several streams that have been offset along the Altyn Tagh fault zone, a major strike-slip fault system that separates the Tibetan Plateau (south) and the Tarim Basin (north). Given the offset streams in the view, what is the sense of offset along this fault?
   a. left-lateral
   b. right-lateral

17. **Strike-Slip Faults - Kazakhstan.** The Problem 17 placemark points to an elliptical-shaped plutonic intrusion in Kazakhstan near Lake Balkhash that has been offset by a fault. Given the offset boundaries of the pale-colored igneous rock in the view, what is the sense of offset along this fault?
   a. left-lateral
   b. right-lateral

Folds are classified by a wide range of names, depending on their geometry: **anticlines**, **synclines**, **domes**, **basins**, and **monoclines** *(look these up in your textbook to learn their characteristics)*. The following questions will take you to locales that provide examples of some of these remarkable structures.

18. **Folds - Zagros Fold Belt, Iran.** The Problem 18 placemark takes you to a spectacular fold in the Zagros Fold Belt of Iran. What type of fold is this? *Recall that **flatirons** commonly form on both tilted **fold limbs** with the tip of the flatiron pointing in the direction opposite of the dip of the fold limb.*
    a. anticline (beds dip <u>away from</u> the central **fold hinge** with oldest rocks at the center)
    b. syncline (beds dip <u>toward</u> the central **fold hinge** with youngest rocks at the center)

19. **Folds - Bolivia.** The Problem 19 placemark takes you to a well-exposed fold in Bolivia. What type of fold is this? *Recall that **flatirons** commonly form on both tilted **fold limbs** with the tip of the flatiron pointing in the direction opposite of the dip of the fold limb.*
    a. anticline (beds dip <u>away from</u> the central **fold hinge** with oldest rocks at the center)
    b. syncline (beds dip <u>toward</u> the central **fold hinge** with youngest rocks at the center)

Folds are said to be **upright** in areas where their fold hinge is approximately horizontal *(bedding contacts tend to be straight and parallel to the main fold hinge)* and **plunging** in areas where their fold hinge is inclined into the ground *(bedding contacts tend to curve and wrap around from one fold limb to another)*. <u>Anticlines plunge in the direction in which the rock layers are convex</u> as they wrap around from one fold limb to another (looking like a stretched-out anticline), whereas <u>synclines plunge in the direction in which the rock layers are concave</u> as they wrap around from one fold limb to another (looking like a stretched-out syncline). Commonly, adjacent anticlines and synclines plunge in the same direction. *To visualize this, fold a piece of paper into an anticline/syncline and plunge it into a table (trace the intersection of the paper and the table to see the map pattern).*

20. **Folds - Morocco.** Study the fold highlighted by the Problem 20 placemark *(you may also want to fly around this area and look at the structure in map and perspective view)*. Which of the following is correct about <u>this portion</u> of this fold?
    a. the fold is not plunging
    b. the fold is plunging to the NW
    c. the fold is plunging to the SE

21. **Folds - Sheep Mountain, WY.** The Problem 21 placemark takes you to Sheep Mountain, WY. Turn on the *Sheep Mt. Geologic Map* overlay. The markers with numbers show the orientation of bedding—the longer line shows the strike direction and the shorter line shows the dip direction. The number indicates the angle of dip *(e.g., the NE fold limb has dips to the NE ranging from the 40s to the 80s, whereas the SW fold limb has dips to the SW from the high 20s to the 40s)*. The map legend identifies the types of fold axes and the colored rock units from oldest (Mm, bottom) to youngest (Kmr, top). Which of the following best describes the area around the Problem 21 placemark?
    a. NW-plunging anticline
    b. NW-plunging syncline
    c. SE-plunging anticline
    d. SE-plunging syncline

22. **Folds - Sheep Mountain, WY.** Using the fold axis and the map pattern of the rock units at the Problem 22 placemark, what is this type of fold and which direction is it plunging?
    a. NW-plunging anticline
    b. NW-plunging syncline
    c. SE-plunging anticline
    d. SE-plunging syncline

23. **Folds - Williamsport, PA.** Which direction is the fold at the Problem 23 placemark plunging and what type of fold is it? *Oftentimes the plunging "noses" of anticlines are long and tapering, whereas the plunging "noses" of synclines are more blunt and rounded.*
   a. WSW-plunging anticline
   b. WSW-plunging syncline
   c. ENE-plunging anticline
   d. ENE-plunging syncline

24. **Folds - Hurricane, UT.** The Problem 24 placemark points to a prominent flatiron of Triassic Shinarump Conglomerate on a fold limb. What name best describes the overall fold? *You might turn on the Geologic Map of Quail Creek SP, UT overlay and make it semi-transparent to assist your interpretation (check out the cross section in the map legend as well).*
   a. SW-plunging anticline
   b. SW-plunging syncline
   c. NE-plunging anticline
   d. NE-plunging syncline

25. **Folds - Hurricane, UT.** Which direction do the beds at the Problem 25 placemark strike?
   a. NE/SW
   b. NW/SE
   c. N/S

26. **Folds - Hurricane, UT.** Which direction do the beds at the Problem 26 placemark dip?
   a. NE
   b. SE
   c. SW
   d. NW

The nearby Circle Cliffs area allows us to explore a fold that is commonly associated with the Colorado Plateau area: the **monocline**. The Circle Cliffs erode into the monocline that forms the Waterpocket Fold of Capitol Reef National Park in Utah.

27. **Folds - Circle Cliffs, UT.** Turn on the *Geologic Map of Circle Cliffs, UT* overlay and locate the flatirons of Jurassic Kayenta (Jk) and Wingate (Jw) on the west and east sides of the Circle Cliffs area (Problem 27a and 27b placemarks highlight two representative flatirons on the different limbs of the fold). Monoclines typically have a steeply tilted fold limb (forelimb) and then a shallowly dipping (sometimes sub-horizontal) backlimb. Which placemark corresponds to the steep forelimb? *You may want to temporarily set the vertical exaggeration to "3".*
   a. Problem 27a placemark
   b. Problem 27b placemark

28. **Folds - Circle Cliffs, UT.** Both monoclines and anticlines uplift and deform rock layers such that when eroded, certain ages of rocks are exposed in the center of the structure relative to the flanking limbs. What is the relative ages of the rocks exposed in the Circle Cliffs area (Problem 28 placemark)? *You may want to use the Map Symbols legend where older rocks are listed at the bottom (Permian formations) with progressively younger units above (Uncons. Quat. deposits).*
   a. older rocks are exposed in the center and younger rocks in the flanking flatirons
   b. younger rocks are exposed in the center and older rocks in the flanking flatirons

29. **Folds - East Kaibab Monocline, AZ.** This stream valley exposes a continuous cross section through the East Kaibab Monocline. Which of the following is correct?
   a. Problem 29a placemark = tilted strata of the monoclinal fold limb
   b. Problem 29b placemark = uplifted sub-horizontal strata within the fold
   c. Problem 29c placemark = relatively undeformed sub-horizontal strata beyond the fold

When rock layers plunge in all directions away from a central area (like the top of a baseball cap), the fold is referred to as a **dome**. Conversely, when rock layers plunge in all directions toward a central area (like a basin in which you might wash your hands), the fold is called a **basin**. Domes and basins are more equidimensional than anticlines and synclines, respectively (that is, doubly-plunging anticlines and synclines have a long fold hinge between their plunging ends). In addition, most cross sections through a dome will look like anticlines, and most cross sections through a basin will look like synclines.

30. **Folds - Bolivia.** Use the Problem 30 placemarks to fly to two different folds in Bolivia. Which of the following is true?
    a. Problem 30a placemark = dome, whereas Problem 30b placemark = anticline
    b. Problem 30a placemark = anticline, whereas Problem 30b placemark = dome
    c. Problem 30a placemark = basin, whereas Problem 30b placemark = syncline
    d. Problem 30a placemark = syncline, whereas Problem 30b placemark = basin

Checkerboard Mesa (37.227°N, 112.880°W) lies within Zion National Park in southern Utah. Here, the massive sand dunes of a once expansive Jurassic desert have been lithified to form the spectacular Navajo Sandstone. The criss-crossing lines that give Checkerboard Mesa its name derive their origin from the sub-horizontal/ scoop-shaped cross-beds created by the shifting sand dunes that intersect the sub-vertical fractures that characterize Zion National Park and the surrounding areas. The cross-beds formed as the sand was deposited, whereas the fractures developed at a much later time after lithification of the rock. Even later, erosion exhumed the artistic landform that we see today.

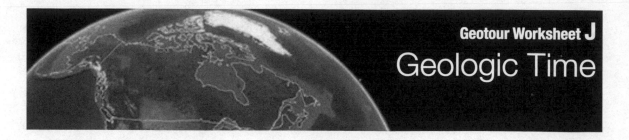

Geotour Worksheet **J**
Geologic Time

Circle the letter next to the best answer for each question.

Worksheet Resources:
- **Google Earth** - Open the **2. Exploring Geology Using Geotours > J. Geologic Time** folder.
  1. **Problem Materials** - **Check and double-click items associated with each problem** to travel to the appropriate location with the prescribed perspective/zoom.
  2. Geologic Time Geotours Library - **Explore additional Geotours** in this folder to help answer problems.
- **Geoscience** - **Consult a textbook** and/or **Internet resources** to help answer some problems.

---

1. **Principle of Original Continuity - Grand Canyon National Park (NP), AZ.** According to the principle of original continuity, which layer corresponds to the layer indicated by the Problem 1 placemark?
   - a. Problem 1a
   - b. Problem 1b
   - c. Problem 1c

2. **Principle of Superposition - Grand Canyon NP, AZ.** According to the principle of superposition, what is the order of age of layering for placemarks Problem 2a (variegated), 2b (pink), and 2c (dark) from oldest to youngest?
   - a. Problem 2a, 2b, 2c
   - b. Problem 2c, 2b, 2a
   - c. Problem 2b, 2c, 2a
   - d. Problem 2a, 2c, 2b

3. **Principle of Cross-Cutting Relationships - Grand Canyon NP, AZ.** According to the principle of cross-cutting relationships, which layer is younger: Problem 3a placemark (pink) or Problem 3b placemark (tan)? *Hint: trace the pink layer to the left and see what happens to it.*
   - a. Problem 3a
   - b. Problem 3b

4. **Unconformities - Grand Canyon NP, AZ.** The surface dividing the units in Problem 3 is called the "Great Unconformity" because it separates older layered Precambrian units from the younger sedimentary Cambrian unit with up to a 1-billion-year gap in the rock record due to erosion. What type of unconformity is it at this location?
   - a. nonconformity
   - b. disconformity
   - c. angular unconformity

5. **Unconformities/Principle of Cross-Cutting Relationships - Grand Canyon NP, AZ.** The Problem 5 placemark highlights another exposure of the Great Unconformity. Here metamorphic rock and a white igneous dike are truncated and overlain by the same tan sedimentary Cambrian unit. Which is older: the dike or the tan Cambrian rock layer?
   - a. dike (Problem 5a placemark)
   - b. tan sedimentary Cambrian rock layer (Problem 5b placemark)

6. **Unconformities - Grand Canyon NP, AZ.** What type of unconformity is the Great Unconformity at this location where sedimentary rocks unconformably overlie crystalline metamorphic and igneous rocks?
   a. nonconformity
   b. disconformity
   c. angular unconformity

7. **Unconformities/Principle of Inclusions - Grand Canyon NP, AZ.** According to the principle of inclusions, which rock unit should have inclusions of the other?
   a. the dike & metamorphic rock should have inclusions of the tan Cambrian rock layer
   b. the tan sedimentary Cambrian rock layer should have inclusions of the dike & metamorphic rock

8. **Unconformities - Grand Canyon NP, AZ.** The Problem 8 placemark points to another unconformity between the Early Mississippian Redwall Limestone (below; steep cliff) and the Early Pennsylvanian Supai Group (above). The unconformity represents a hiatus of about 25 million years. What type of unconformity is this? *Hint: You know that both units are sedimentary. From the view you have, determine if the layers above and below the unconformity are parallel.*
   a. nonconformity
   b. disconformity
   c. angular unconformity

Just for Fun...Turn on the *Grand Canyon Geologic Map* overlay and explore the region in Google Earth (perhaps first look at the *Stratigraphic Overview* placemark). Not only can you find multiple examples of the principles used to determine relative dating, but the map also shows you what geologic periods these units have been correlated with (which have been dated using absolute age-dating techniques). *Note that the map is 4 MB in size and will take a moment to load.*

The Grand Canyon provides spectacular exposures of Paleozoic and Precambrian rock layers. As you travel north from the Grand Canyon, you encounter laterally extensive, younger Mesozoic rock layers that have eroded into a series of cliffs and benches. Because of this stair-like erosional profile caused by differences in rock type, the area is known as the "Grand Staircase". The Moenave Formation forms most of the "Vermillion Cliffs", the Navajo Sandstone forms the "White Cliffs", and the Claron Formation forms the "Pink Cliffs".

9. **Mesozoic Stratigraphy - Grand Staircase National Monument (NM), UT.** Turn on the *Geologic Map of Grand Staircase NM, UT* overlay and make it semi-transparent. Check and double-click the Problem 9 placemarks. Which placemark corresponds to the correct cliff-forming unit? *Use the "Map Symbols" section of the map to assist you. You might want to temporarily change the vertical exaggeration to "3" to see the "steps" of the Grand Staircase.*
   a. Problem 9a: Moenave; Problem 9b: Navajo; and Problem 9c: Claron
   b. Problem 9a: Claron; Problem 9b: Navajo; and Problem 9c: Moenave
   c. Problem 9a: Navajo; Problem 9b: Claron; and Problem 9c: Moenave
   d. Problem 9a: Moenave; Problem 9b: Claron; and Problem 9c: Navajo

10. **Mesozoic Stratigraphy - Grand Staircase NM, UT.** The steep cliffs of Zion Canyon at Zion National Park in Utah compose one of the "steps" in the Grand Staircase. Check and double-click the Problem 10 placemark, and turn on the *Geologic Map of Zion NP* overlay to determine which cliff forms Zion Canyon.
    a. White Cliffs
    b. Vermillion Cliffs
    c. Pink Cliffs

11. **Mesozoic Stratigraphy - Grand Staircase NM, UT.** Checkerboard Mesa, one of the most famous landmarks in Zion National Park, is composed of this same cliff-forming unit. Check and double-click the Problem 11 folder to fly there. The "checkerboard" pattern is due to the intersection of sandstone cross beds (yellow) with joints (green). Which of these formed <u>first</u> (one of them formed while the rock was sediment)?

      a. joints

      b. cross-beds

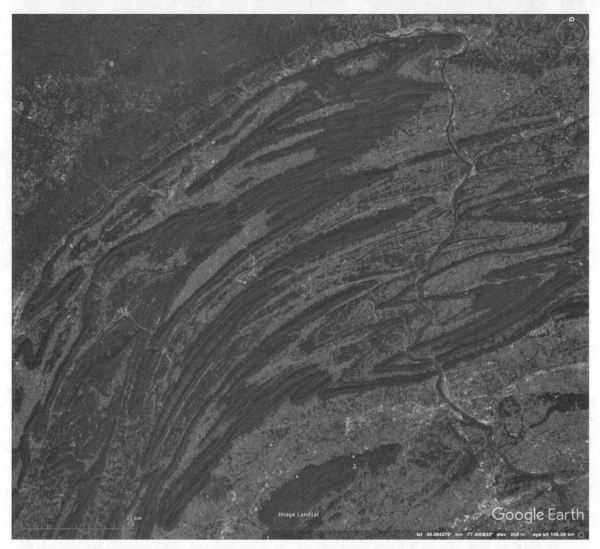

From high above the Earth, the contorted folds of the central Appalachian Mountains in Pennsylvania (40.654°N, 077.524°W) hint of a long and complex deformational history. From this vantage point, however, it is impossible to unravel that this region records three distinct Paleozoic orogenies and a late Paleozoic–Early Mesozoic rifting event that led to the formation of the Atlantic Ocean. Such details require a thorough investigation of the depositional and deformational histories of the rocks preserved in the area.

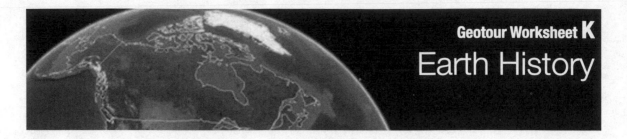

Geotour Worksheet K
Earth History

Circle the letter next to the best answer for each question.

Worksheet Resources:
- **Google Earth** - Open the **2. Exploring Geology Using Geotours > K. Earth History** folder.
  1. **Problem Materials** - **Check and double-click items associated with each problem** to travel to the appropriate location with the prescribed perspective/zoom.
  2. **Earth History Geotours Library** - **Explore additional Geotours** in this folder to help answer problems.
- **Geoscience** - **Consult a textbook** and/or **Internet resources** to help answer some problems.

Check the *Paleogeographic Maps* folder in the *Global Paleogeographic Model* folder and make the entire folder semi-transparent. Turn on the latitude and longitude lines *(View > Grid)* and *Borders and Labels* in the *Layers* panel. Now, play the time animation through several times and rotate the globe to watch the animation from different perspectives. You can manually play the animation by grabbing the right part of the time slider and moving it, or you can toggle the animation on/off automatically by clicking the right-most clock icon in the historical animation box (upper left corner of the Google Earth viewer). Leave the *Paleogeographic Maps* folder checked to answer the following questions.

1. **Earth History/Paleogeography - World.** Check and double-click the Problem 1 placemark. Approximately 600 Ma (make sure that the time slider in the upper left corner of the Google Earth viewer window is all the way to the left and that the 601 Ma label is visible in the lower left hand corner), the continents were combined into one huge supercontinent that was mostly over what present-day ocean?
     a. Atlantic
     b. Pacific
     c. Indian
     d. Arctic
2. **Earth History/Paleogeography - World.** Check and double-click the Problem 2 placemark. Approximately 460-470 Ma, were the landmasses predominantly in the northern or southern hemisphere?
     a. northern
     b. southern
3. **Earth History/Paleogeography - World.** Check and double-click the Problem 3 placemark. In the early Paleozoic, vast areas of the continents were flooded with shallow seas called epicontinental seas (420-430 Ma, for example) where life flourished. How are these areas characterized on the paleogeographic maps?
     a. brown areas
     b. dark blue areas
     c. light blue areas

4. **Earth History/Paleogeography - World.** Check and double-click the Problem 4 placemark. Watch the landmass in the center of the view as the animation goes from the Problem 3 to Problem 4 placemark (do this a couple of times to study it carefully). This landmass will become North America. During this time period, it is growing by accreting terranes to its margins. What kind of plate tectonic boundary is likely responsible for these series of mountain-building events?
    a. transform
    b. divergent
    c. convergent

5. **Earth History/Paleogeography - World.** Check and double-click the Problem 5 placemark. A major continent-continent collision is occurring to form what supercontinent at around 270-280 Ma?
    a. Rodinia
    b. Pangaea
    c. Laurentia
    d. Gondwana

6. **Earth History/Paleogeography - World.** Check and double-click the Problem 6 placemark. What mountain belt does this convergence form (note that the eastern coast of North America figures prominently in this collision)?
    a. Himalayas
    b. Ediacarans
    c. Mazatzals
    d. Appalachians/Caledonides

7. **Earth History/Paleogeography - World.** Check and double-click the Problem 7 placemark. When did the supercontinent start rifting apart?
    a. 350-300 Ma
    b. 280-250 Ma
    c. 220-170 Ma
    d. 100-50 Ma

8. **Earth History/Paleogeography - World.** Check and double-click the Problem 8 placemark. When did South America and Africa begin to separate?
    a. 170-105 Ma
    b. 220-180 Ma
    c. 100-50 Ma
    d. 50-25 Ma

9. **Earth History/Paleogeography - World.** Check and double-click the Problem 9 placemark. Approximately how long did it take India to travel from Antarctica to collide with Asia? Start at 120 Ma and progressively drag the time slider until India collides with Asia.
    a. ~5 million years
    b. ~30 million years
    c. ~15 million years
    d. ~90 million years

Geotour Worksheet **L**

Energy & Mineral Resources

The Bingham Canyon/Kennecott Copper Mine (40.524°N, 112.150°W) in the Oquirrh Mountains of north-central Utah is a massive testimony to the extent that humans will go to acquire mineral resources. Excavation of this large porphyry copper deposit has created the largest human-produced excavation in the world, encompassing an area of nearly 8 km² (2000 acres).

Circle the letter next to the best answer for each question.

Worksheet Resources:
- **Google Earth** - Open the **2. Exploring Geology Using Geotours > L. Energy & Mineral Resources** folder.
  1. **Problem Materials** - **Check and double-click items associated with each problem** to travel to the appropriate location with the prescribed perspective/zoom.
  2. **Energy & Mineral Resources Geotours Library** - **Explore additional Geotours** in this folder to help answer problems.
- **Geoscience** - **Consult a textbook** and/or **Internet resources** to help answer some problems.

Check the *Major Known Oil Reserves* folder and <u>make the entire folder semi-transparent</u>. Also, turn on *Borders and Labels* in the *Layers* panel.

1. **Hydrocarbon Resources - World.** Which of the following regions is <u>not</u> a location that has major oil reserves?
   a. Gulf of Mexico
   b. Saudi Arabia
   c. Japan
   d. Niger Delta, Africa

2. **Hydrocarbon Resources - World.** Which continent currently does <u>not</u> have any major known oil reserves?
   a. Asia
   b. Africa
   c. Antarctica
   d. South America

3. **Hydrocarbon Resources - Offshore United States.** Movement of subsurface salt deposits can produce complex structures, including **salt domes** and **salt anticlines**. When this happens, the salt that flows into domes must be withdrawn from somewhere else, causing the overlying beds to sink down to form an adjacent **basin**. Hydrocarbons sometimes migrate from these basins to become trapped in reservoirs adjacent to the margins of domes. Check and double-click the Problem 3a and 3b placemarks. Which feature does each placemark point to?
   a. Problem 3a: basin and Problem 3b: dome/anticline
   b. Problem 3b: basin and Problem 3a: dome/anticline

4. **Hydrocarbon Resources - Offshore United States.** Check and double-click the Problem 4 placemark. What kind of hydrocarbon-trapping structure is this?
   a. dome
   b. basin

5. **Hydrocarbon Resources - Offshore United States.** To get a sense of scale of this feature, measure its diameter using the *Ruler* tool (in km between the Problem 5 placemarks). What is its diameter?
   a. 5-15 km
   b. 20-30 km
   c. 0.1-0.5 km
   d. 0.01-0.05 km

Name:_____

6. **Coal Resources - Farmersburg, IN.** Coal mine reclamation commonly involves refilling the active pit of the surface coal mine after coal has been extracted, spreading topsoil over the area, and then replanting it. Despite these efforts, reclaimed land may not look like unmined land for quite some time (if ever). Which placemark identifies a region that has been reclaimed in this manner in 2005?
   a. Problem 6a
   b. Problem 6b
   c. Problem 6c

7. **Coal Resources - Farmersburg, IN.** Extracting coal in open-pit strip mines may lead to unwanted consequences. When water reacts with minerals in excavated rock, the runoff into lakes and streams contains elements that affect water quality. Check and double-click the Problem 7 placemark (also fly over the mine and nearby reclaimed areas). What evidence from 2005 do you see that suggests possible contamination (which is much improved in present-day imagery)?
   a. water in the ponds has an unusual green color
   b. orange sediments suggest iron oxidation
   c. both a & b
   d. neither a nor b

8. **Coal Resources - Centralia, PA.** The Problem 8 placemark flies you to an area just north of the former town of Centralia, PA, which has the infamous distinction of being home to an underground coal fire that has burned since 1962. The fire is believed to have been started by burning trash in a former strip mine, leading to a coal seam catching on fire. Coal seam traces (e.g., water-filled area near Aristes, PA highlighted by the Problem 8 placemark) can readily be traced on the surface, providing an explanation for how the coal seam could catch on fire from burning trash. Which of the following is the best explanation for how the coal seam provides fuel for the underground fire? *Hint: Look carefully at the Problem 8 placemark area.*
   a. the coal seam is essentially flat and is only exposed in valleys by erosion
   b. the coal seam is essentially flat, but has been exposed by strip mining in some areas and remains underground in other areas
   c. the coal seam has actually been completely strip-mined away, but piles of rocks representing the former overburden that covered the coal seam now cover tailings that still have burnable coal that has caught fire
   d. the coal seam is actually folded into anticlines and synclines with some parts of the seam intersecting the surface and other parts remaining in the subsurface

In addition to hydrocarbons and coal, there are other types of energy sources as well.

9. **Other Energy Resources - Rockford, IL.** Check and double-click the Problem 9 placemark to fly to an area south of Rockford, IL. What type of energy source is shown here?
   a. wind farm
   b. hydroelectric
   c. solar farm
   d. geothermal

10. **Other Energy Resources - Rockford, IL.** Is this type of energy "green" (renewable, nonpolluting) or "not green" (it uses nonrenewable resources and is possibly polluting the environment)?
    a. green
    b. not green

11. **Other Energy Resources - Rockford, IL.** Check and double-click the Problem 11 placemark to fly to an area NW of Rockford, IL. What type of energy source is shown here?
    a. wind farm
    b. hydroelectric
    c. nuclear
    d. solar farm

12. **Other Energy Resources - Rockford, IL.** Check and double-click the Problem 12 placemark to zoom out from this area to an altitude of about 250 km. Assuming that weather systems follow a general path from west to east, what major metropolitan area is downwind of this facility?
    a. Madison, WI
    b. Milwaukee, WI
    c. Quad Cities (Rock Island, IL; Moline, IL; Davenport, IA; and Bettendorf, IA)
    d. Chicago, IL

As we saw in the C. Minerals worksheet, mineral resources that we use in many everyday items must be extracted from the Earth as well. Let's take a closer look at the example from the title page of this worksheet.

13. **Mineral Resources - Bingham Canyon/Kennecott Copper Mine, UT.** Check and double-click the Problem 13 placemarks to fly to Bingham Canyon/Kennecott Copper Mine in Utah, the largest open-pit mine in the world. Using the *Ruler* tool (in km between the Problem 13 placemarks), estimate the diameter of the main pit.
    a. 11-15 km
    b. 3-5 km
    c. 0-1 km
    d. 21-25 km

14. **Mineral Resources - Bingham Canyon/Kennecott Copper Mine, UT.** Check and double-click the Problem 14 placemarks and use the *Hand* cursor to measure the elevation difference (in m) between the two placemarks to get an estimate for the depth of the main pit
    a. 900-1000 m
    b. 1200-1300 m
    c. 250-350 m
    d. 80-140 m

15. **Mineral Resources - Bingham Canyon/Kennecott Copper Mine, UT.** Check and double-click the Problem 15 placemark to visit one of the many tailings/waste piles. What evidence exists in this view from 2012 to suggest that this process has similar environmental issues to the coal strip mine in Indiana?
    a. water ponds have discolored water
    b. reclaimed areas have only small trees on them

Name:_____

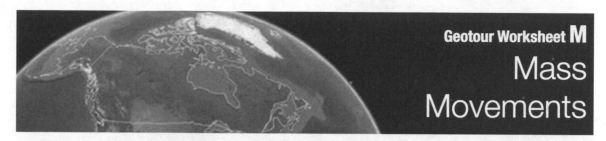

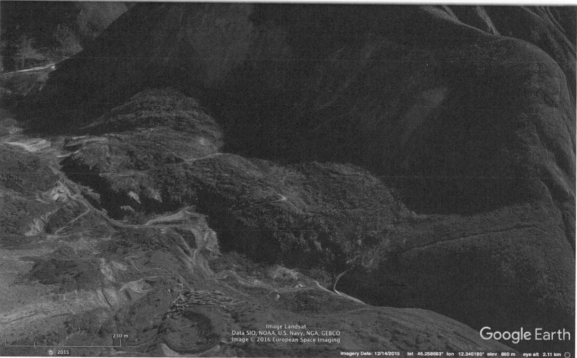

The north slope of Monte Toc (46.266°N, 012.339°E) gave way on October 9, 1963 in a massive landslide that collapsed into a reservoir formed by the Vajont Dam (lower right portion of the figure). The resulting 250-meter-high wall of water overtopped the dam and flowed south down the valley, destroying many villages and towns downstream.

Name:_____

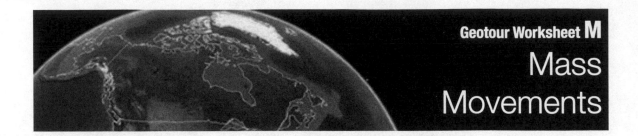

Circle the letter next to the best answer for each question.

Worksheet Resources:
- **Google Earth** - Open the **2. Exploring Geology Using Geotours > M. Mass Movements** folder.
  1. **Problem Materials** - **Check and double-click items associated with each problem** to travel to the appropriate location with the prescribed perspective/zoom.
  2. **Mass Movements Geotours Library** - **Explore additional Geotours** in this folder to help answer problems.
- **Geoscience** - **Consult a textbook** and/or **Internet resources** to help answer some problems.

---

1. **Landslide - Portuguese Bend, CA.** The Problem 1 placemark flies you to the Portuguese Bend area in California. This region has experienced a massive (and progressive) landslide that has cannibalized the subdivision that once existed on the cliffs overlooking the ocean. Many of these buildings reside on a hummocky debris apron. How did urbanization likely contribute to cause the mass wasting? *You may want to use your textbook.*
   a. water from septic systems, sprinklers, etc. lubricated the slope making it more unstable
   b. undercutting the slope to make flatter areas for buildings made the slope more unstable
   c. weight of buildings and roads made the area more unstable
   d. all of the above

2. **Landslide - Portuguese Bend, CA.** Do you think making the road at the location marked by the Problem 2 placemark was a smart thing to do?
   a. yes, it will help stabilize the slope
   b. yes, it will serve as a boundary to prevent further building on the unstable area
   c. no, the road just undercuts the slope above it, making it even more unstable

3. **Landslide - Portuguese Bend, CA.** How has this homeowner attempted to address slope stability?
   a. installed an irrigation system to make the soil more cohesive
   b. planted rows of trees to retard soil movement
   c. placed rock on the slope to slow erosion

4. **Mudslide - La Conchita, CA.** The Problem 4 placemark takes you to an ENE view of a steep, seaward-sloping face behind the flat, coastal sea terrace on which people have built homes. The placemark is positioned on the most recent, active mudslide, which is the infamous La Conchita mudslide. Why do you think this mudslide occurred?
   a. the area has little vegetation to hold the soil
   b. the area could have been undercut by the road halfway up the slope
   c. the sea terrace on which the houses sit has over-steepened the cliff
   d. all of the above

5. **Mudslide - La Conchita, CA.** What do the areas marked by the Problem 5 placemarks represent?
   a. construction sites
   b. older, smaller mudslides
   c. former tsunami erosional features

6. **Mudslide - La Conchita, CA.** Check and double-click the Problem 6 placemarks in sequence to view the La Conchita mudslide area in 1994, 2003, 2005, and present-day in Street View. What happened in 1995 that foreshadows the major 2005 La Conchita mudslide?
    a. a mudslide that destroyed a few homes
    b. creep and minor downslope movement
    c. major road building along the slope above the town

7. **Landslide - Gros Ventre Slide, WY.** The Problem 7 placemark takes you to the Gros Ventre Slide that formed Slide Lake. To what specific part of the slide is the placemark pointing?
    a. a scar at the "head" of the slide where trees and soil have been stripped away
    b. a hummocky debris apron of material at the "toe" of the slide
    c. the location has not experienced any mass movement

8. **Landslide - Gros Ventre Slide, WY.** Debris from the landslide dammed up the Gros Ventre River to form Slide Lake. Based on the position of the lake, in what direction does the river flow at this locality?
    a. west to east
    b. east to west

9. **Landslide - Gros Ventre Slide, WY.** Which is not a possible reason that might have contributed to the development of the slide?
    a. river erosion may have made the slope unstable
    b. heavy rains may have made the materials heavier and lubricated the slip surface
    c. rock layers were dipping toward the valley, parallel to the slip surface
    d. a large housing development was built in the area, making the slope unstable

10. **Landslide - Vajont Dam Slide, Italy.** These placemarks fly you to the site of the Vajont Dam disaster that occurred in Italy in 1963. Which answer correctly identifies the features highlighted by the Problem 10a, 10b, and 10c placemarks?
    a. Problem 10a: dam; Problem 10b: hummocky debris apron; and Problem 10c: slide scar
    b. Problem 10a: hummocky debris apron; Problem 10b: slide scar; and Problem 10c: dam
    c. Problem 10a: slide scar; Problem 10b: dam; and Problem 10c: hummocky debris apron

11. **Landslide - Vajont Dam Slide, Italy.** Which is not a possible reason that might have contributed to the development of the slide?
    a. the layers dip toward the valley
    b. water from the reservoir formed by the dam lubricated the slip surface
    c. the weight of buildings in the Italian village of Casso (located at the Problem 11 placemark)

12. **Rockslide - Frank Slide, Alberta, Canada.** In 1903, limestone from Turtle Mountain slid downslope, engulfing part of the mining town of Frank. Check and double-click the Problem 12 placemark to look south along the length of Turtle Mountain. Based on your analysis from this viewpoint, which of the following is one of the likely causes of the rockslide?
    a. the east limb of the Turtle Mountain anticline dips steeply to overturned here
    b. water from nearby Crowsnest Lake lubricated the slip surface
    c. major urban development on the west limb of the anticline created pressure on the east limb due to the weight of the buildings and water from their septic fields

13. **Slumps - Velarde, NM.** At this locale, slumps on the SE face of Black Mesa are sliding into the Rio Grande River valley. The slumps extend the land surface here, effectively producing what type of fault in which the hanging wall (slipped material) moves down with respect to the footwall (unslipped material)?
    a. reverse fault
    b. normal fault
    c. strike-slip fault

14. **Landslide Remediation - Hiroshima, Japan.** Check and double-click the Problem 14 placemarks to look at this mountainside in Hiroshima over time (turn off *Historical Imagery* for the present-day view). How have engineers attempted to mitigate future landslides?
    a. they have channeled landslide materials into diversion ditches that flow into the ocean
    b. they have replanted landslide areas to stabilize the slopes
    c. they have placed barriers across existing landslides

Name:_____

Goosenecks State Park (37.162°N, 109.937°W) near Mexican Hat, UT provides spectacular examples of entrenched meanders. Here, the San Juan River once meandered as a low gradient stream on its way to join the Colorado River. With uplift of the Colorado Plateau, opening of the Gulf of California, and increased discharge due to stream piracy, the Colorado River rapidly downcut its channel, lowering the local base level for the San Juan River. In response to these abrupt changes, the San Juan River incised into its own floodplain without experiencing meander cutoffs, forming the deep, meandering canyons that we see today.

Circle the letter next to the best answer for each question.

Worksheet Resources:
- **Google Earth** - Open the **2. Exploring Geology Using Geotours > N. Stream Landscapes** folder.
  1. **Problem Materials** - **Check and double-click items associated with each problem** to travel to the appropriate location with the prescribed perspective/zoom.
  2. Stream Landscapes Geotours Library - **Explore additional Geotours** in this folder to help answer problems.
- **Geoscience** - **Consult a textbook** and/or **Internet resources** to help answer some problems.

---

1. **Headward Erosion - Canyonlands National Park, UT.** In this view from Grand View Point in Canyonlands National Park, three stream branches (the middle one is labeled with the placemark) are experiencing **headward erosion**. Headward erosion allows streams to lengthen in the upstream direction. As seen from this viewpoint, in which direction (right or left) are the streams lengthening?
    a. right (west)
    b. left (east)

2. **Headward Erosion - Montrose, CO.** In which direction are the streams experiencing headward erosion?
    a. NE
    b. SW
    c. NW
    d. SE

Streams can develop a variety of patterns depending on the rock type and structure that underlies them. See your textbook for diagrams of stream patterns.

3. **Stream Patterns - Namibia.** What is the stream pattern exhibited at the Problem 3 placemark?
    a. trellis *(streams meet at right angles)*
    b. rectangular *(streams have right-angle bends)*
    c. radial *(streams radiate from a central high point)*
    d. dendritic *(streams resemble veins in a leaf)*

4. **Stream Patterns - Mifflintown, TN.** Check and double-click the Problem 4 folder. What is the overall general stream pattern exhibited at this location with tributaries (thin blue lines) flowing into the main trunk stream (thick blue line)? *Hint: Pay particularly close attention to the tributaries flowing from Shade Mountain (SE side). Note that some streams (e.g., the stream highlighted in cyan) exhibit a rectangular drainage pattern as they follow intersecting fracture sets and/or resistant rock layers.*
    a. trellis *(streams meet at right angles)*
    b. parallel *(streams are approximately parallel to each other)*
    c. radial *(streams radiate from a central high point)*
    d. dendritic *(streams resemble veins in a leaf)*

5. **Stream Patterns - Novaya Zemlya, Russia.** Check and double-click the Problem 5 folder. Segments of several streams have been highlighted to emphasize where the streams follow different intersecting fracture sets. What stream pattern is exhibited in this area?
   a. parallel *(streams are approximately parallel to each other)*
   b. rectangular *(streams have right-angle bends)*
   c. radial *(streams radiate from a central high point)*
   d. dendritic *(streams resemble veins in a leaf)*

6. **Stream Patterns - Chile.** What is the stream pattern exhibited at the Problem 6 placemark?
   a. trellis *(streams meet at right angles)*
   b. parallel *(streams are approximately parallel to each other)*
   c. rectangular *(streams have right-angle bends)*
   d. radial *(streams radiate from a central high point)*

7. **Stream Patterns - Mount Merapi, Indonesia.** What is the stream pattern exhibited at the Problem 7 placemark?
   a. trellis *(streams meet at right angles)*
   b. parallel *(streams are approximately parallel to each other)*
   c. rectangular *(streams have right-angle bends)*
   d. radial *(streams radiate from a central high point)*

8. **Meandering Streams - Rio Ucayali River, Peru.** What is the outside bend of a meander called (Problem 8 placemark)?
   a. cut bank
   b. point bar
   c. oxbow lake
   d. yazoo tributary

9. **Meandering Streams - Rio Ucayali River, Peru.** What is the deposit on the inside bend of a meander called (Problem 9 placemark)?
   a. cut bank
   b. point bar
   c. natural levee
   d. meander scar

10. **Meandering Streams - Rio Ucayali River, Peru.** Check and double-click the Problem 10 placemark to fly to the Pucallpa, Peru area. Watch the animated GIF image in the placemark, or click on the link in the placemark balloon for the Google Earth Engine website for this same region. After viewing the animation several times, which of the following statements is _not_ true?
    a. meanders migrate to form meander cutoffs
    b. meanders migrate laterally
    c. meanders migrate upstream
    d. meanders migrate downstream

Just for Fun...Revisit the Google Earth Engine view for Problem 10 and look at not only how the river's course changes relative to the city of Pucallpa (top center), but also the growth of Pucallpa and the extent of deforestation over this time period (expansion of light green versus dark green).

11. **Meandering Streams - Vicksburg, MS.** What is the stream mostly flowing in the larger river's floodplain called (Problem 11 placemark)? *Be sure to turn Borders and Labels in the Layers panel.*
    a. point bar
    b. oxbow lake
    c. yazoo tributary
    d. meander scar

12. **Meandering Streams - Natchez, LA.** What is an abandoned meander with water called (Problem 12 placemark)?
   a. cut bank
   b. oxbow lake
   c. yazoo tributary
   d. natural levee

13. **Meandering Streams - Goosenecks State Park, UT.** As you have seen, most meandering streams occupy broad, flat floodplains. What has happened to the meandering stream shown by the Problem 13 placemark?
   a. it has become incised/entrenched because of a change in gradient or base level
   b. it has carved a steep canyon because it experienced a catastrophic flood
   c. it is a normal meandering stream that has just experienced a lot of mass movements

14. **Meandering Streams - Bowknot Bend Area, UT.** Check and double-click the Problem 14 placemark. This meander has been cut off sometime in the past and now has been bypassed by the Green River. Did the cutoff happen before or after the river incised/entrenched into the landscape?
   a. after, because the bend has been incised/entrenched as well
   b. before, the river could not have cut through the meander neck if it was already entrenched

15. **Stream Piracy - Kaaterskill, NY.** Check and double-click the Problem 15 folder. The dark blue and orange streams highlight the process of **stream piracy/stream capture** (i.e., they are pirating/capturing the headwaters of the cyan and magenta streams, respectively). Which of the following helps promote this?
   a. the pirating streams are larger and can hold more water than the other streams
   b. the pirating streams are eroding headwardly to intersect more of the other streams' drainage basins, causing water to be diverted down their steeper gradients
   c. the land to the west is being uplifted
   d. landslides have diverted the water into the stream pirates

16. **Discordant Streams - Bighorn River, Sheep Mountain, WY.** Check and double-click the Problem 16 placemark. Notice that the stream cuts across the Sheep Mountain anticline instead of being diverted around it. Such **discordant** streams are generally called **antecedent streams** or **superposed streams**. Antecedent streams are generally older than the structure and continue their downcutting at a pace similar to the rate at which uplift occurs (such areas commonly experience present-day earthquakes). In contrast, superposed streams develop long after a structure has developed and has been buried by sediment. The younger stream develops its stream valley and cuts through the older structure as it exhumes it. In the *Layers* panel, turn on *Gallery > Earthquakes*. What is the most likely interpretation for the stream?
   a. antecedent
   b. superposed

17. **Discordant Streams - Arun River, Himalaya Mountains, Nepal.** Check and double-click the Problem 17 placemark, and leave *Layers > Gallery > Earthquakes* turned on. Trace the Arun River to its headwaters. What is the most likely interpretation for this river that cuts across the Himalaya mountains?
   a. antecedent
   b. superposed

18. **Catastrophic Flooding - Katun River, Russia.** Check and double-click the Problem 18 placemarks to fly to the Altai Republic in Russia. Here, a catastrophic flood formed mega-ripples when a dam impounding glacial meltwaters burst. To get a sense of scale of these mega-ripples, use the *Ruler* tool to measure (in m) between the ripple crests marked by the Problem 18 placemarks.
   a. ~10 m
   b. ~50 m
   c. ~100 m

Name:_____

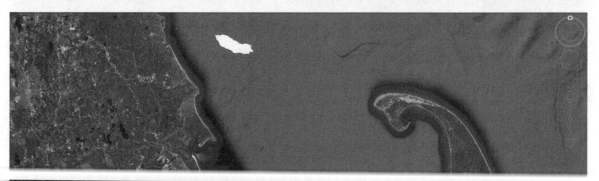

Image Landsat
Data SIO, NOAA, U.S. Navy, NGA, GEBCO
Image U.S. Geological Survey

Google Earth

Imagery Date: 6/23/2016   lat 41.781735° lon -70.416595° elev -19 m   eye alt 91.63 km

The shoreline forming the Cape Cod peninsula (41.680°N, 070.246°W) is easily recognizable by its distinctive "muscle-flexing arm" shape. The "shoulder", "biceps", and "forearm" portions consist of sand and gravel remnants of glacial moraines that were deposited some 18,000 years ago. These sand and gravels were subsequently reworked by currents to form the "elbow" and "curled fist" sections of Cape Cod.

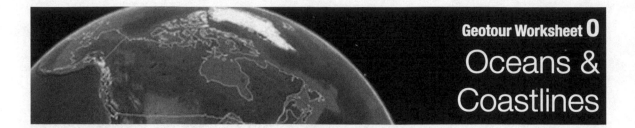

Geotour Worksheet O

Oceans & Coastlines

Circle the letter next to the best answer for each question.

Worksheet Resources:
- **Google Earth** - Open the **2. Exploring Geology Using Geotours > O. Oceans & Coastlines** folder.
  1. **Problem Materials** - **Check and double-click items associated with each problem** to travel to the appropriate location with the prescribed perspective/zoom.
  2. Ocean & Coastlines Geotours Library - **Explore additional Geotours** in this folder to help answer problems.
- **Geoscience** - **Consult a textbook** and/or **Internet resources** to help answer some problems.

---

1. **Seafloor Bathymetry - Argentine Sea, South America.** Check and double-click the Problem 1a, 1b, and 1c placemarks to fly to the southern tip of South America. Identify the bathymetric feature associated with each placemark. *See your textbook if you forget what these things are.*
   a. Prob 1a: continental shelf; Prob 1b: continental slope; and Prob 1c: abyssal plain
   b. Prob 1a: continental slope; Prob 1b: continental shelf; and Prob 1c: abyssal plain
   c. Prob 1a: abyssal plain; Prob 1b: continental slope; and Prob 1c: continental shelf
   d. Prob 1a: abyssal plain; Prob 1b: continental shelf; and Prob 1c: continental slope

2. **Seafloor Bathymetry - South Atlantic Ocean.** Check and double-click the Problem 2 folder. Identify the bathymetric expression of the tectonic plate boundary associated with each placemark. *See your textbook if you forget what these things are.*
   a. Prob 2a: trench; Prob 2b: mid-ocean ridge; and Prob 2c: transform fault
   b. Prob 2a: mid-ocean ridge; Prob 2b: trench; and Prob 2c: transform fault
   c. Prob 2a: mid-ocean ridge; Prob 2b: transform fault; and Prob 2c: trench
   d. Prob 2a: transform fault; Prob 2b: trench; and Prob 2c: mid-ocean ridge

3. **Seafloor Bathymetry - South Atlantic Ocean.** Check and double-click the Problem 3a and 3b placemarks. Which placemark shows bathymetry associated with a **passive margin** (no nearby active plate boundary)? *Note that the other placemark highlights an **active margin** (associated with an active plate boundary).*
   a. Problem 3a
   b. Problem 3b

Coral reefs are communities of organisms that can build spectacular landforms in relatively shallow water depths (<60 m) at low latitudes (<30°).

4. **Coral Reefs - Tahaa, South Pacific Ocean.** Check and double-click the Problem 4 placemark to fly to some scenic islands in the South Pacific. What kind of coral reef is surrounding the island?
   a. barrier reef *(forms offshore from the coast with an intervening lagoon)*
   b. fringing reef *(forms directly along the coast)*
   c. atoll *(forms a circular reef surrounding a lagoon)*

5. **Coral Reefs - Tahiti, South Pacific Ocean.** Check and double-click the Problem 5 placemark to fly to a nearby island in the South Pacific. What kind of coral reef is surrounding the island?
   a. barrier reef *(forms offshore from the coast with an intervening lagoon)*
   b. fringing reef *(forms directly along the coast)*
   c. atoll *(forms a circular reef surrounding a lagoon)*

6. **Coral Reefs - Maldives, Indian Ocean.** Check and double-click the Problem 6 placemark to fly to the Maldives in the Indian Ocean. What kind of coral reef is present here?
    a. barrier reef *(forms offshore from the coast with an intervening lagoon)*
    b. fringing reef *(forms directly along the coast)*
    c. atoll *(forms a circular reef surrounding a lagoon)*

Barrier islands, beaches, and spits form in coastal areas with abundant sand.

7. **Barrier Islands - Cape Hatteras, NC.** Check and double-click the Problem 7 placemark to fly to Cape Hatteras off the coast of North Carolina. Use the *Hand* tool to determine the range of heights for the sand ridge specified by the placemark *(presumably this sand ridge would provide protection from waves of this height or lower for areas on the lagoon side of the ridge)*.
    a. 0-1 m
    b. 3-12 m
    c. 15-25 m
8. **Barrier Islands - Cape Hatteras, NC.** Why is most of the construction on the lagoon side of the barrier island (Problem 8 placemark)?
    a. more sunlight
    b. protected from storm waves and erosion
    c. land there is at a higher elevation
    d. there are more beach sands on that side
9. **Barrier Islands - Chincoteague Bay, MD.** Check and double-click the Problem 9 placemark to fly to a location along Chincoteague Bay, MD. Watch the animated GIF image in the placemark, or click on the link in the placemark balloon for the Google Earth Engine website for this same region. After viewing the animation several times, which of the following statements is <u>not</u> true?
    a. barrier islands do not experience mobilization/transport of sand
    b. gaps in the barrier island can be foci for erosion
    c. gaps in the barrier island can create distributary channels that both transport and deposit sand on the calmer lagoon side of the barrier island
10. **Spits - Cape Cod, MA.** Check and double-click the Problem 10 placemarks to fly to the arm-shaped coastline of Cape Cod, MA. Here, the remains of an E-W oriented glacial moraine form the "muscle" of the "arm", and a N-S glacial moraine creates most of the N-S "forearm". **Spits** are being created at the curled "fist" near Provincetown, MA (Problem 10a placemark) and the "elbow" near Chatham, MA (Problem 10b placemark). Watch the animated GIF image in the placemarks, or click on the link in the placemark balloons for the Google Earth Engine website for these same regions. Determine the general direction of the current at these locations from these animations *(look at which way material is moving)*?
    a. Problem 10a placemark - WNW and Problem 10b placemark - NNE
    b. Problem 10a placemark - ESE and Problem 10b placemark - SSW
    c. Problem 10a placemark - WNW and Problem 10b placemark - SSW
    d. Problem 10a placemark - ESE and Problem 10b placemark - NNE
11. **Beach Preservation - Chicago, IL.** Piers or groins have been built out into Lake Michigan in Chicago to retain sand for beaches and to prevent sand erosion (Problem 11 placemark). Which direction is the current that is moving the sand? *The current direction is from the narrow part of the sand "triangle" toward the wider part.*
    a. from N to S
    b. from S to N

<u>Just for Fun</u>...Check out the Google Earth Engine view for Problem 11 to check your answer.

12. **Sea Level Changes - East Coast, USA.** Turn on the *U.S. East Coast Sea Level Changes* overlay (make it semi-transparent), and double-click the Problems 12a, 12b, 12c, and 12d placemarks. If the Greenland and Antarctica ice sheets melt, where would be the best place to own land?
    a. Problem 12a
    b. Problem 12b
    c. Problem 12c
    d. Problem 12d

Groundwater & Karst Landscapes

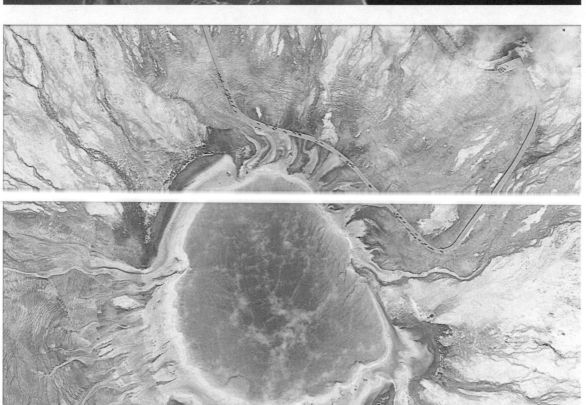

Yellowstone National Park in Wyoming contains one of the world's greatest concentrations of geothermal hot springs and geysers, fueled by groundwater heated by the underlying Yellowstone Hot Spot. Grand Prismatic Spring (44.525°N, 110.838°W) provides a spectacular example of Yellowstone's hot springs (note the boardwalk and people for scale). The pool derives its vibrant colors from thermophile bacteria – dark shades of blue indicating the regions of hottest water, whereas the more muted yellows and browns reflect areas of cooler water.

Geotour Worksheet **P**

Groundwater &
Karst Landscapes

Circle the letter next to the best answer for each question.

Worksheet Resources:
- **Google Earth** - Open the **2. Exploring Geology Using Geotours > P. Groundwater & Karst Landscapes** folder.
  1. **Problem Materials** - **Check and double-click items associated with each problem** to travel to the appropriate location with the prescribed perspective/zoom.
  2. Groundwater & Karst Landscapes Geotours Library - **Explore additional Geotours** in this folder to help answer problems.
- **Geoscience** - **Consult a textbook** and/or **Internet resources** to help answer some problems.

---

1. **Groundwater Reserves & Irrigation - Saudi Arabia.** Check and double-click the Problem 1 placemark to fly to a location in Saudi Arabia. Watch the animated GIF image in the placemark, or click on the link in the placemark balloon for the Google Earth Engine website for this same region. After viewing the animation several times, which of the following statements is <u>not</u> true?
   a. groundwater wells are providing the source of water for the circular irrigation pivots
   b. the general trend of the irrigated areas is NE-SW
   c. there is an overall increase in the number of irrigation wells over time

2. **Groundwater Reserves & Irrigation - Saudi Arabia.** Given the regional trend of irrigation wells (Problem 2 path, assume that the middle section is covered by thick sand deposits), what is the best explanation for this trend and why these irrigation wells are only located in certain "bands"?
   a. they just haven't gotten around to drilling the other areas
   b. the water may be contained in **aquifers** that have this trend
   c. the other areas receive plenty of surface water and don't require well water

Surface water flow is influenced most directly by elevation differences. Similarly, **groundwater flow** is controlled by differences in elevation of the **water table**, which is the surface between completely saturated and partially saturated porous rocks/sediments. These water table elevation differences often reflect surface elevations...to a degree.

3. **Groundwater (and Surface Water) Flow - Everglades National Park (NP), FL.** Use the *Hand* tool to determine the elevations at the Problem 3a and 3b placemarks. What direction should the surface and groundwater flow? *(Confirm this by turning on the Everglades, FL-Previous Groundwater Flow overlay).*
   a. N to S
   b. S to N
   c. E to W

4. **Groundwater (and Surface Water) Flow - Everglades NP, FL.** Turn on the *Everglades, FL-Present Groundwater Flow* overlay. What has changed?
   a. some swamps have been drained for urban areas, altering the groundwater flow
   b. canals have diverted water from Lake Okeechobee to coastal cities
   c. groundwater flow directions have been modified by the canal system
   d. all three of the above answers are correct

5. **Groundwater (and Surface Water) Flow - Everglades NP, FL.** In some areas of Florida, particularly near the coast, various urbanization changes (e.g., pumping freshwater from wells) has altered the water table such that it slopes landward instead of seaward. Why is this a problem? *Keep the Everglades, FL-Present Groundwater Flow overlay turned on.*
   a. salt water is less dense than freshwater, so wells will have to be drilled deeper to tap freshwater resources
   b. salt water is more dense than freshwater, causing dissolution of limestone and the formation of sinkholes
   c. saltwater incursion/intrusion will contaminate freshwater aquifers because the groundwater flow direction has been modified by the change in water table slope
   d. this will cause massive erosion along coastal areas, potentially destroying many of Florida's spectacular beaches

Yellowstone National Park in Wyoming is home to many spectacular geothermal **hot springs** and **geysers**, whose groundwater heat source is derived from the underlying Yellowstone Hot Spot.

6. **Hot Springs - Grand Prismatic Spring, Yellowstone NP, WY.** The Problem 6 placemarks fly you to Midway Geyser Basin to see Grand Prismatic Spring (a hot spring) and the nearby Firehole River in Yellowstone NP. Use the *Ruler* tool to measure the width of Grand Prismatic Spring (in m)

   a. 125-135 m
   b. 291-301 m
   c. 266-276 m
   d. 980-990 m

7. **Hot Springs - Grand Prismatic Spring, Yellowstone NP, WY.** Turn on the Problem 7 placemarks. Using the *Hand* tool, compare elevations of Grand Prismatic Spring and the Firehole River (in m). Does overland flow of water travel from the spring to the Firehole River or from the Firehole River to Grand Prismatic Spring?
   a. Firehole River to Grand Prismatic Spring
   b. Grand Prismatic Spring to Firehole River

8. **Hot Springs - Mammoth Hot Springs, Yellowstone NP, WY.** Check and double-click the Problem 8 placemark to fly to Mammoth Hot Springs, an area in the NW corner of Yellowstone NP. This area is different than the silica-based geyser features to the south that are associated with volcanic rocks. Here, hot groundwater circulates in the subsurface to <u>dissolve minerals from the Madison Limestone</u>. What are these terraces of re-precipitated minerals made of *(recall what mineral typically forms limestone)*?
   a. quartz
   b. calcite
   c. gypsum
   d. pyrite

Acidic groundwater can dissolve soluble rocks to form fascinating **karst landscapes**.

9. **Karst Features - Bosnia and Herzegovina.** Check and double-click the Problem 9 placemark to fly to an area in west-central Bosnia and Herzegovina (part of the former country of Yugoslavia, where karst landscapes were first described). What are the depressions that dot the landscape called (sometimes they can be filled with water or clumps of trees)? *Note: These features are very similar to those found in the Pennyroyal Plateau area just outside Mammoth Cave National Park, KY. You can visit this area using placemarks in the Groundwater & Karst Landscapes Geotours Library.*
    a. dry holes
    b. springs
    c. sinkholes
    d. caves

10. **Karst Features - Orleans, IN.** Check and double-click the Problem 10 path to fly to an area near Orleans, IN. This path highlights part of the Lost River, a stream that flows from east to west in this area. Follow the stream path east to west, and look at the water level in the stream and at the landscape through which it flows. What kind of stream is it?
    a. intermittent
    b. disappearing/sinking
    c. arroyo
    d. perennial

11. **Karst Features - Orleans, IN.** The Lost River really isn't "lost". West of the Problem 10 path, we see glimpses of the water that has been diverted underground in areas like Wesley Chapel Gulf (Problem 11a placemark, also referred to as Elrod Gulf). Such areas are called **karst windows**, as they provide a "window" into the groundwater system. Eventually, the groundwater resurfaces at **resurgent springs** near Orangeville *(The Problem 11b placemark highlights the Orangeville Rise. Another resurgent spring, Rise of Lost River, occurs 1.2 km (0.75 mi) to the SW at the Problem 11c placemark.)*, where the Lost River flows at the surface again. Why does the Lost River flow at the surface after it reappears at the Orangeville Rise/Rise of Lost River?
    a. the rugged landscape is covered by insoluble sandstone and not by soluble limestone
    b. the Lost River actually goes back into the subsurface just downstream of the Rise of Lost River

12. **Karst Features - Longdeng, Guangxi, China.** What kind of karst landscape is shown in the area around the Problem 12 placemark?
    a. tower karst
    b. sinkhole plain
    c. karst window

13. **Karst Features - Winter Park, FL.** What kind of karst feature is shown by the Problem 13 placemark (see photo in the placemark)?
    a. cave
    b. monadnock
    c. dry valley
    d. sinkhole

14. **Karst Features - Arecibo Observatory, Puerto Rico.** In what kind of karst feature does the radio telescope likely reside?
    a. cave
    b. monadnock
    c. dry valley
    d. sinkhole

Name:_____

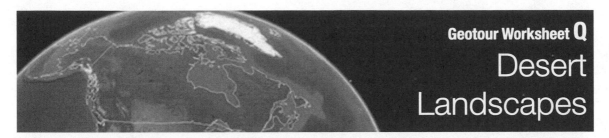

Desert Landscapes

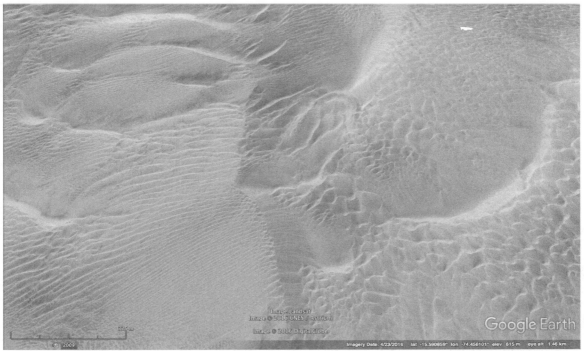

Sand dunes in the Atacama Desert (15.588°S, 074.456°W) on the west coast of Peru support the contention that this region is the driest non-polar place on Earth. According to the USGS, coastal deserts like the Atacama are commonly found on the western margins of continental land masses near the Tropic of Cancer or the Tropic of Capricorn (the latter in this instance) where measurable rainfall is extremely infrequent.

Geotour Worksheet **Q**
Desert Landscapes

Circle the letter next to the best answer for each question.

Worksheet Resources:
- **Google Earth** - Open the **2. Exploring Geology Using Geotours > Q. Desert Landscapes** folder.
  1. **Problem Materials** - **Check and double-click items associated with each problem** to travel to the appropriate location with the prescribed perspective/zoom.
  2. Desert Landscapes Geotours Library - **Explore additional Geotours** in this folder to help answer problems.
- **Geoscience** - **Consult a textbook** and/or **Internet resources** to help answer some problems.

A **desert** is a region that is so arid that it contains no permanent streams, except for those that bring water in from temperate regions elsewhere.

1. **Deserts - Cairo, Egypt.** The Problem 1 placemark flies you to the Pyramids of Giza on the southwestern outskirts of Cairo, Egypt. These structures are located on the eastern edge of the largest desert in the world (the desert covers most of northern Africa). What is the name of this desert?
   - a. Gobi
   - b. Sahara
   - c. Mojave
   - d. Kalahari

2. **Deserts - Tanaca, Peru.** Which desert is the driest place on Earth? Check and double-click the Problem 2 placemark to fly to this desert along the western coast of South America.
   - a. Gobi
   - b. Sahara
   - c. Atacama
   - d. Mojave

3. **Deserts - Taklimakan Desert, Tarim Basin, China.** Check and double-click the Problem 3 polygon to fly to the Tarim Basin in China. Use the *Ruler* tool to estimate the longest axis of the Tarim Basin polygon (km).
   - a. 200-300 km
   - b. 900-1000 km
   - c. 50-150 km
   - d. 1200-1400 km

4. **Deserts - Taklimakan Desert, Tarim Basin, China.** Why does this north-flowing stream terminate in the middle of the desert, even though the land slopes gently north?
   - a. it disappears into a sinkhole
   - b. evaporation and infiltration are too great
   - c. it has been dammed up to create a reservoir

5. **Desert Features - Guerro Negro, Baja, Mexico.** Check and double-click the Problem 5 placemark to fly to a location in Baja, Mexico. Watch the animated GIF image in the placemark, or click on the link in the placemark balloon for the Google Earth Engine website for this same region. <u>Toward what direction did the wind blow</u> to create these dunes? *Hint: The tips of these dunes form downwind, making the dunes concave in the downwind direction.*
   a. SE
   b. NW
   c. NE
   d. SW

6. **Desert Features - Ayers Rock, Australia.** Check and double-click the placemarks for Problem 6 to fly to Ayers Rock in Australia. Use the *Hand* tool to determine how high this rock stands above its surroundings (in m, measure the placemarks).
   a. 500-530 m
   b. 90-130 m
   c. 300-350 m
   d. 800-850 m

7. **Desert Features - Ayers Rock, Australia.** Ayers Rock is an erosional remnant of one limb of a regional syncline (the lines across the rock are steeply dipping beds). What term is used to describe such resistant erosional remnants that stand out in relief above the alluvium in arid
   a. klippe
   b. roche moutonnée
   c. inselberg
   d. horn

8. **Desert Features - Death Valley, CA.** The Problem 8 placemark highlights a common feature that forms evaporite deposits in arid settings like Death Valley. What is the highlighted feature?
   a. playa lake
   b. bajada
   c. butte
   d. inselberg

9. **Desert Features - Death Valley, CA.** The Problem 9 placemarks highlight depositional features that commonly occur in arid settings where there is an abrupt change in stream/valley gradient. What are the highlighted features?
   a. alluvial fans
   b. mesas
   c. buttes
   d. inselbergs

10. **Desert Features - Joshua Tree National Park, CA.** The Problem 10 placemark highlights an area where the depositional features in Problem 9 have coalesced along a mountain front to form a larger, compound feature. What is this highlighted feature called?
    a. bajada
    b. mesa
    c. playa lake
    d. inselberg

11. **Desert Features - Monument Valley, AZ.** The Problem 11 placemarks highlight erosional features that commonly occur in arid settings. **Buttes** tend to have a flat top with a small surface area relative to their height, whereas **mesas** are flat-topped with a large surface area relative to their height. What are the highlighted features?
    a. Problem 11a - mesa; Problem 11b - butte
    b. Problem 11a - butte; Problem 11b - mesa

Water usage in arid regions is becoming an increasingly important issue, especially in large metropolitan areas like Phoenix, AZ.

12. **Water Usage in Arid Regions - Phoenix, AZ.** The Problem 12 placemarks highlight two distinctly different areas near Phoenix, AZ...one natural to the region and one artificially supported by irrigation. Which of the following is correct?
   a. Problem 12a - lush, green golf course; Problem 12b - arid desert with dry washes
   b. Problem 12a - arid desert with dry washes; Problem 12b - lush, green golf course

13. **Water Usage in Arid Regions - Phoenix, AZ.** How, in part, are all of the lush yards/golf courses and the large human population supported? *Turn on the Problem 13 folder to see all seven!*
   a. landslides
   b. marinas
   c. dams
   d. wildlife sanctuaries

14. **Water Usage in Arid Regions - Phoenix, AZ.** What is the feature highlighted by the Problem 14 placemark?
   a. interstate
   b. water canal
   c. railroad
   d. bike path

15. **Water Usage in Arid Regions - Phoenix, AZ.** Follow the feature in Problem 14 to the west (continue across the tunnels/pipelines that look like gaps). Where does it ultimately originate *(the Problem 15 placemarks highlight the ends of a long tunnel)*?
   a. Colorado River
   b. Pacific Ocean
   c. a glacier in the Sierra Nevada mountains
   d. Gulf of California

Just for Fun...Considering your answers to the questions above, are Phoenix and the surrounding communities **sustainable**? Can they grow larger? Should they grow larger?

16. **Water Usage in Arid Regions - Aral Sea, Central Asia.** Check and double-click the Problem 16 placemark in the Problem 16 folder to fly to the Aral Sea in Central Asia to study what has happened as rivers that once flowed into the Aral Sea have been diverted for agricultural irrigation. Watch the animated GIF image in the placemark, or click on the link in the placemark balloon for the Google Earth Engine website for this same region. Over this time period, what has happened to the size/surface area of the Aral Sea? *Check and double-click the Ships I and Ships II placemarks to zoom into a portion of Zhalanash Harbor to see effects of these changes.*
   a. the Aral Sea has experienced less than ~10% reduction in surface area
   b. the Aral Sea has experienced between ~10% to 50% reduction in surface area
   c. the Aral Sea has experienced greater than ~50% reduction in surface area

Name:_____

The jagged tooth-like peak of the Matterhorn (45.976°N, 007.659°E) dominates this glaciated vista of the Swiss Alps. The Matterhorn is, in fact, a glaciated "horn" (a peak that has been carved by glaciers on three or more sides). In the foreground at the base of the Matterhorn, two glaciers merge into a single glacier with distinctive lateral moraines forming ridges along its margins.

101

Circle the letter next to the best answer for each question.

Worksheet Resources:
- **Google Earth** - Open the **2. Exploring Geology Using Geotours > R. Glacial Landscapes** folder.
    1. **Problem Materials** - **Check and double-click items associated with each problem** to travel to the appropriate location with the prescribed perspective/zoom.
    2. Glacial Landscapes Geotours Library - **Explore additional Geotours** in this folder to help answer problems.
- **Geoscience** - **Consult a textbook** and/or **Internet resources** to help answer some problems.

Continental glaciers created a wide array of fascinating landforms (many of which people pass by every day without noticing).

1. **Continental Glaciers - Palmyra, NY.** Check and double-click the Problem 1 placemark to fly to northern New York State. Once there, temporarily increase your vertical exaggeration to 3. Now click the *Google Maps* button in the icon bar. Once Google Maps opens in your web browser, click the *Menu* icon (upper left-hand corner) and select *Terrain* from the menu (you may have to toggle *Satellite* off before *Terrain* becomes selectable), which allows you to see a shaded relief map with **contour** lines (lines of equal elevation). What are the numerous landforms that you see here?
    a. kettles
    b. drumlins/flutes

2. **Continental Glaciers - Palmyra, NY.** Using the geometry of these features, <u>from what direction</u> did the continental glacier advance (originate)? Glaciers generally advanced from the direction where the contour lines on the landforms are closest together (steep slope) toward the direction where the contour lines on the landforms are farther apart (gentle slope).
    a. N
    b. S

3. **Continental Glaciers - Palmyra, NY.** What are these features mostly composed of? *You may need to refer to your textbook to determine what they are quarrying.*
    a. ice-carved rock
    b. glacial till
    c. tightly folded anticlines & synclines

4. **Continental Glaciers - Finger Lakes Region, NY.** Check and double-click the Problem 4 placemark to see the nearby Finger Lakes region of northern New York State. How did these lakes form?
    a. these were subglacial tunnels that channeled meltwater
    b. these formed from the melting of icebergs
    c. glaciers scoured deep grooves in the landscape that filled with water

5. **Continental Glaciers - Long Island, NY & Cape Cod, MA.** Check and double-click the Problem 5 map overlay. What kind of moraines were deposited on Long Island & the E-W part of Cape Cod?
    a. recessional/terminal/end
    b. lateral
    c. medial
    d. ground

6. **Continental Glaciers - Long Island, NY & Cape Cod, MA.** Given your answer to Problem 5, <u>from what direction</u> did the continental glacier advance?
   a. N
   b. S
   c. E
   d. W

7. **Continental Glaciers - Iverson, MN.** Check and double-click the Problem 7 folder to fly to northeastern Minnesota in order to see two different end/recessional moraines left by continental glaciers. Once there, temporarily increase your vertical exaggeration to 3. Zoom into the Problem 7a and 7b placemarks to see the **kame and kettle topography** of one of the moraines. Which placemark highlights a kame *(note that the other is a kettle)*?
   a. Problem 7a placemark
   b. Problem 7b placemark

8. **Continental Glaciers - Iverson, MN.** The Problem 8 placemark highlights a sinuous ridge that is approximately perpendicular to the end/recessional moraine. Such ridges actually formed under the glacial ice and commonly contain accumulations of gravel and sand *(note the quarry on the NW end of the ridge)*. What is this glacial featured called?
   a. lateral moraine
   b. kame

   d. kettle

9. **Continental Glaciers - Iverson, MN.** Which of the following is <u>not</u> true?
   a. Problem 9a placemark is a kettle
   b. Problem 9b placemark is an esker
   c. Problem 9c placemark is a kame
   d. Problem 9d placemark is a drumlin

Alpine and valley glaciers sculpt the landscape over which they flow and create numerous spectacular landforms, most commonly in mountainous settings. The following problems will explore some of these in more detail *(please reset your vertical exaggeration to 1)*.

10. **Alpine/Valley Glaciers - Baffin Island, Canada.** Check and double-click the Problem 10 placemarks to fly to Baffin Island in Canada. What type of moraine do these placemarks highlight?
    a. ground
    b. lateral
    c. medial
    d. terminal/end

11. **Alpine/Valley Glaciers - Baffin Island, Canada.** What type of moraine does the Problem 11 placemark highlight?
    a. ground
    b. lateral
    c. medial
    d. terminal/end

12. **Alpine/Valley Glaciers - Baffin Island, Canada.** The glacier highlighted by the Problem 12 placemark extends down the valley to actually have its terminus extend into the seawater of the fjord. What type of glacier is this?
    a. tidewater
    b. piedmont
    c. cirque

13. **Alpine/Valley Glaciers - Swiss Alps, Switzerland.** Check and double-click the Problem 13 placemarks to fly to the Matterhorn area in Switzerland. What narrow, knife-like feature that once separated two alpine glaciers do these placemarks highlight?
    a. cirque
    b. horn
    c. truncated spur
    d. arête

14. **Alpine/Valley Glaciers - Swiss Alps, Switzerland.** Check and double-click the Problem 14 placemarks. What semi-circular, amphitheater-shaped feature do these placemarks highlight?
    a. cirque
    b. horn
    c. truncated spur
    d. arête

15. **Alpine/Valley Glaciers - Swiss Alps, Switzerland.** Check and double-click the Problem 15 placemarks. What jagged feature that is a result of glaciers sculpting the rock on three or more sides do these placemarks highlight?
    a. cirque
    b. horn
    c. truncated spur
    d. arête

16. **Alpine/Valley Glaciers - Swiss Alps, Switzerland.** Check and double-click the Problem 16 placemark. What saddle-like feature along the knife-like ridge does this placemark highlight?
    a. tarn
    b. col on an arête
    c. truncated spur
    d. hanging valley

17. **Alpine/Valley Glaciers - Glacier National Park (NP), MT.** Check and double-click the Problem 17 placemark to fly to the Garden Wall in Glacier NP. What feature does the Garden Wall represent?
    a. cirque
    b. horn
    c. col on an arête
    d. arête

18. **Alpine/Valley Glaciers - Glacier NP, MT.** Check and double-click the Problem 18 placemark. What feature does this placemark highlight?
    a. cirque
    b. tarn
    c. col on an arête
    d. truncated spur

19. **Alpine/Valley Glaciers - Rocky Mountain NP, CO.** Check and double-click the Problem 19 placemark. What feature does this placemark highlight?
    a. cirque
    b. tarn
    c. hanging valley

20. **Alpine/Valley Glaciers - Canadian Rocky Mountains, Canada.** Check and double-click the Problem 20 placemark to fly to the Columbia Ice Field in the Canadian Rocky Mountains. Watch the animated GIF image in the placemark, or click on the link in the placemark balloon for the Google Earth Engine website for this same region. What is the rate (m per year) that the Columbia Glacier receded from 1984 to 2004 *(measure the distance between placemarks)*?
    a. 10-25 m per year
    b. 35-80 m per year
    c. 90-120 m per year
    d. 130-160 m per year

21. **Alpine/Valley Glaciers - Holy Cross, CO.** Check and double-click the Problem 21 placemark to fly to Holy Cross, CO. What feature dams the end of this glacially carved valley?
    a. terminal/end moraine
    b. lateral moraine
    c. medial moraine
    d. arête

22. **Alpine/Valley Glaciers - Mono Lake, CA.** Check and double-click the Problem 22 folder to fly to Mono Lake, CA. All of the placemarks highlight the same type of feature. What are these features?
    a. terminal/end moraines
    b. lateral moraines
    c. ground moraines
    d. medial moraines

23. **Alpine/Valley Glaciers - Mono Lake, CA.** Check and double-click the placemarks for Problems 23a and 23b. Using cross-cutting relationships, which moraine is the youngest? *Recall that younger features truncate and cut across older features.*
    a. Problem 23a
    b. Problem 23b

placemark to fly to the Saint Elias Mountains in Alaska. Watch the animated GIF image in the placemark, or click on the link in the placemark balloon for the Google Earth Engine website for this same region. What direction(s) are glaciers flowing in the NW-SE valley?
    a. NW & SE
    b. NW
    c. SE

25. **Alpine/Valley Glaciers - Tokositna & Ruth Glaciers, Denali NP, AK.** Check and double-click the Problem 25 placemark to fly to the southern flank of Mount McKinley in Denali NP. Watch the animated GIF image in the placemark, or click on the link in the placemark balloon for the Google Earth Engine website for this same region. The time-lapse animation shows multiple medial moraines of Tokositna (west) and Ruth (east) Glaciers. Why do the moraines show multiple colors?
    a. they are covered by different amounts of snow
    b. they are eroded from different rock types
    c. they have weathered differently
    d. it is an optical effect due to the position of the moraines relative to the Sun

Sometimes alpine and valley glaciers spill out from an ice field/cap and coalesce when the topography becomes less steep, forming **piedmont glaciers** and associated landforms.

26. **Piedmont Glaciers - Malaspina Glacier, AK.** Check and double-click the Problem 26 placemark to fly to the Malaspina Glacier area in Alaska. What are the folds composed of?
    a. layers of moraine and ice
    b. layers of solid rock

27. **Piedmont Glaciers - Malaspina Glacier, AK.** Turn on the Problem 27 path to show the approximate boundary between ice from the piedmont glacier and rock and/or ice from a nearby glacier. What is causing the layers to be bent & distorted?
    a. plate tectonic convergence
    b. movement along a fault
    c. frictional drag as the glacial ice flows past the rocks to the right/east of the Problem 27 path
    d. this distortion is only apparent, the materials were actually deposited this way

28. **Piedmont Glaciers - Malaspina Glacier, AK.** Check and double-click the Problem 28 placemark. What circular water-filled depression does this placemark highlight?
   a. kettle
   b. kame

29. **Piedmont Glaciers - Malaspina Glacier, AK.** Check and double-click the Problem 29 placemark. What feature does this placemark highlight?
   a. braided stream on an outwash plain
   b. esker

Name:_____

The encroachment of deforestation on the Amazon rain forest in Brazil (09.235°S, 062.915°W). Light-colored areas depict regions where the rain forest has been clear-cut by human activity, whereas the dark green areas remain less disturbed by humans. Deforestation often is preceded by the construction of roads and subsequently expands from road margins into the rain forest.

Circle the letter next to the best answer for each question.

Worksheet Resources:
- **Google Earth** - Open the **2. Exploring Geology Using Geotours > S. Global Change** folder.
  1. **Problem Materials** - **Check and double-click items associated with each problem** to travel to the appropriate location with the prescribed perspective/zoom.
  2. Global Change Geotours Library - **Explore additional Geotours** in this folder to help answer problems.
- **Geoscience** - **Consult a textbook** and/or **Internet resources** to help answer some problems.

One of the current global concerns is the breakup and melting of glaciers and ice shelves due to **global warming**.

1. **Global Warming - Larsen B Ice Shelf, Antarctica.** Check and double-click the Problem 1 placemarks in sequential order (Problem 1a - Jan 31, 2002 *[the dark spots on the ice are pools of water]*, Problem 1b - Mar 07, 2002, & Problem 1c - Feb 23, 2006) to view historical imagery of the Larsen B ice shelf in Antarctica. Much like the "canary in the coal mine", the Larsen B ice shelf has responded to global warming in dramatic fashion (which surprised scientists). What happened? *Note: The Dec 1999 imagery is currently incorrect at the time of this publication.*
   a. the ice shelf disintegrated
   b. the ice shelf dramatically expanded

2. **Global Warming - Larsen B Ice Shelf, Antarctica.** To get a sense of scale, use the *Polygon* tab on the *Ruler* tool to estimate the area of the largest iceberg (in km^2) in the Feb 23, 2006 imagery.
   a. 10-40 km^2
   b. 60-100 km^2
   c. 600-650 km^2
   d. 1000-1200 km^2

Humans also are modifying their environment in various ways (including deforestation) and dramatically affecting both local and global ecosystems.

3. **Deforestation - Amazon Rain Forest, Brazil.** Check and double-click the Problem 3 placemark to fly to the Amazon rain forest in Brazil. Watch the animated GIF image in the placemark, or click on the link in the placemark balloon for the Google Earth Engine website for this same region. Estimate the percentage of land in the field of view that has experienced deforestation from 1984 to 2012.
   a. 1%
   b. 15%
   c. 75%

4. **Deforestation - Amazon Rain Forest, Brazil.** Turn on the Problem 4 polygon and make it semi-transparent. This polygon represents an area of approximately 5000 km². Using your answer from Problem 3, what area (in km²) within the polygon has been deforested from 1984 to 2012?
    a. 500 km²
    b. 1250 km²
    c. 3750 km²
    d. 4750 km²

5. **Deforestation - Amazon Rain Forest, Brazil.** On average, based on your answer to Problem 4, how much deforestation occurred in this area per year (km²/yr)?
    a. 15 km²/yr
    b. 38 km²/yr
    c. 101 km²/yr
    d. 134 km²/yr

Just for Fun...Explore other areas in the Amazon using the Google Earth Engine website to see the impact that humans are having on the rain forest due to deforestation. *This Just for Fun activity is only "fun" in that there are no questions...the results are actually quite sobering.*

Water usage in arid regions is becoming an increasingly important issue related to global change, especially in large metropolitan areas like Phoenix, AZ. *Note: This series of questions duplicates material in the Desert Landscapes worksheet, but is appropriate for this topic as well (especially if you did not do the Desert Landscapes worksheet).*

6. **Water Usage in Arid Regions - Phoenix, AZ.** The Problem 6 placemarks highlight two distinctly different areas near Phoenix, AZ...one natural to the region and one artificially supported by irrigation. Which of the following is correct?
    a. Problem 6a - lush, green golf course; Problem 6b - arid desert with dry washes
    b. Problem 6a - arid desert with dry washes; Problem 6b - lush, green golf course

7. **Water Usage in Arid Regions - Phoenix, AZ.** How, in part, are all of the lush yards/golf courses and the large human population supported? *Turn on the Problem 7 folder to see all seven!*
    a. landslides
    b. marinas
    c. dams
    d. wildlife sanctuaries

8. **Water Usage in Arid Regions - Phoenix, AZ.** What is the feature highlighted by the Problem 8 placemark?
    a. interstate
    b. water canal
    c. railroad
    d. bike path

9. **Water Usage in Arid Regions - Phoenix, AZ.** Follow the feature in Problem 8 to the west (continue across the tunnels/pipelines that look like gaps). Where does it ultimately originate *(the placemarks for Problem 9 mark the ends of a long tunnel)*?
    a. Colorado River
    b. Pacific Ocean
    c. a glacier in the Sierra Nevada mountains
    d. Gulf of California

Just for Fun...Considering your answers to the questions above, are Phoenix and the surrounding communities **sustainable**? Can they grow larger? Should they grow larger?

10. **Preservation of Habitats - Everglades National Park (NP), FL.** Check and double-click the Problem 10 placemark to fly to the Everglades in Florida. In the *Layers* panel turn on *More > Parks / Recreation Areas > US National Parks* to see the green outline of the boundary of the protected national park *(note the sharpness of the eastern boundary with developed areas)*. Based on the alignment of hammocks of grass (the elongate ridges that create a grain to the Everglades), what is the direction of regional water flow in the Everglades?
    a. NNE to SSW
    b. NNW to SSE

11. **Preservation of Habitats - Everglades NP, FL.** North of the park border, if developers or farmers drain and utilize this land, what will happen to the source of water for the Everglades, and therefore to the ecosystem of the Everglades?
    a. the Everglades will be unchanged
    b. the Everglades will become significantly drier

Global warming can be a contributing factor to **changes in sea level** around the world. Turn on the *World Sea Level Trends* folder to answer the following questions. *Note that the <u>direction of placemark arrows</u> indicates whether sea level is rising or falling and that the <u>color of the placemark arrow/label</u> indicates the magnitude of sea level change (see legend).*

12. **Sea Level Changes - United States.** Which coast has locations that have experienced the largest change in sea level (red)?
    a. East Coast (Problem 12a placemark)
    b. West Coast (Problem 12b placemark)
    c. Gulf Coast (Problem 12c placemark)

13. **Sea Level Changes - United States.** For those areas with the largest change in sea level (red), why do you think sea level increases are so high?
    a. a crustal gravity anomaly there creates a larger pull on the water, causing it to flow here
    b. meltwater from nearby glaciers and ice sheets contributes to sea level rise
    c. this area is a delta with sediment that compacts over time, causing the elevation of the land to subside in addition to the rise in sea level

14. **Sea Level Changes - Norway, Sweden, & Finland.** How is sea level changing along the coasts of Norway, Sweden, & Finland (e.g., Gulf of Bothnia)?
    a. sea level is rising
    b. sea level is falling

15. **Sea Level Changes - Norway, Sweden, & Finland.** What is causing this change in sea level?
    a. global warming
    b. subsidence of the land
    c. being closer to the poles
    d. glacial rebound due to isostatic adjustment of the lithosphere in response to the removal of the weight of glacial ice over the area

16. **Sea Level Changes - Alaska.** The Alaskan coastline adjacent to the Gulf of Alaska is experiencing similar sea level changes to Norway, Sweden, & Finland. However, the magnitude of the sea level changes along the Alaskan coastline are much more pronounced. The changes in sea level not only reflect what is happening in Norway, Sweden, & Finland, but also what other contributing factor?
    a. seawater is displaced from this region because it is flowing toward other regions (like water sloshing in a bathtub)
    b. uplift of the land due to a convergent plate boundary
    c. nearby rifting causing a deep trench/rift valley into which seawater is flowing
    d. pumping of hydrocarbons from beneath Prudhoe Bay

17. **Sea Level Changes - New York City, NY.** Turn on *Layers > 3D Buildings > Photorealistic,* and fly to Liberty Island using the Problem 17 placemark. Uncheck the *World Sea Level Trends* folder, and check the *Climate Central Sea Level Layers* folder to generate polygons that represent sea level predictions in response to increases of 1.5°C, 2°C, 3°C, & 4°C in average global temperature. Show/hide the polygons for each temperature level by opening the *Climate Central Sea Level Layers* folder, unchecking the *Historic carbon pollution* folder, and toggling the respective temperature sub-folders on/off *(note that you may need to zoom out and then zoom back in to refresh the polygons).* Which of the following is true?
    a. Global warming of 1.5°C doesn't have a noticeable effect on Liberty Island
    b. Global warming of 2°C floods Liberty Island except for the Statue of Liberty and the area immediately adjacent to its star-shaped base
    c. Global warming of 3°C floods Liberty Island and overtops the Statue of Liberty's star-shaped base
    d. Global warming of 4°C completely submerges Liberty Island and the Statue of Liberty

18. **Sea Level Changes - New York City, NY.** Leave *Layers > 3D Buildings > Photorealistic* turned on, and fly to the 3.4 m (11 ft) tall "Wall Street Bull" statue in New York City using the Problem 18 placemark. Embedded in the placemark is a slider showing the effects of a 2°C and a 4°C rise in temperatures on sea level in the area around the "Wall Street Bull" *(note that the base of the bull is ~7 m (~21 ft) above present-day sea level).* Move the slider back and forth in the placemark, and then select the answer that best describes what you see (elevations are in m).
    a. 2°C - water would be ~7 m higher than present-day (bull's hooves in water), whereas 4°C - water would be ~10.5 m higher than present-day (covering the bull)
    b. 2°C - water would be ~4 m higher than present-day (down the street), whereas 4°C - water would be ~10 m higher than present-day (bull's tail is above water)
    c. 2°C - water would be ~10 m higher than present-day (bull's tail is above water), whereas 4°C - water would be ~15 m higher than present-day (covering the bull)
    d. 2°C - water would be ~2 m higher than present-day (don't see the water), whereas 4°C - water would be ~7 m higher than present-day (bull's hooves in water)

19. **Sea Level Changes - Sydney, Australia.** Leave *Layers > 3D Buildings > Photorealistic* turned on, and fly to the famed opera house in Sydney, Australia using the Problem 19 folder. Embedded in the Problem 19 folder is a slider showing the effects of a 2°C and a 4°C rise in temperatures on sea level in the area. Based on information provided by the interactive slider, which area remains unflooded after an increase in global warming of 4°C?
    a. Problem 19a placemark
    b. Problem 19b placemark

Just for Fun...Explore other coastal areas around the world using the *Climate Central Sea Level Layers* folder to generate polygons that represent sea level predictions in response to increases of 1.5°C, 2°C, 3°C, & 4°C in average global temperature. *This Just for Fun activity is only "fun" in that there are no questions...the results are actually quite sobering. See http://sealevel.climatecentral.org/ maps for more interactive maps involving climate change and sea level.*

© 2016 Google
US Dept of State Geographer
Data SIO, NOAA, U.S. Navy, NGA, GEBCO
Image Landsat

Google Earth

Imagery Date: 12/14/2015   lat 31.890450° lon -97.827285° elev 307 m   eye alt 8076.86 km

In order for users to see and use personal image files (e.g., photos, diagrams, maps, etc.) that you have embedded in your Google Earth project, you have to make the files publicly accessible from a web server rather than simply storing the files on your hard disk (otherwise users cannot see your images and usually will see only a box with a red "X" in it). There are many websites that will allow you to access images online at little to no cost. Basically, all you need is a URL for your image, so it can be viewed in a standard web browser. Below we describe how to access images using a free Dropbox Basic account.

1. Using your web browser, go to the following website:
   http://www.dropbox.com

2. If you already have an account, **Sign in** to that account from the Dropbox home page. If you want to **create a new Dropbox Basic account with 2 GB of free space**, follow the online instructions for creating a new account (you also will need to read and agree to Dropbox's terms). As part of the account setup, Dropbox also will take you through the steps of downloading and installing the Dropbox application for your particular device.

3. To upload a file to use in Google Earth, click the **Upload icon** (page with +) to find and upload the image (e.g., a JPG photo, an animated GIF image, etc.) from your device to the Home folder using your system's standard file dialog box. Once the file is uploaded, others can now view the file in Google Earth (or in web browsers) because Dropbox serves as a web server.

4. To obtain the URL for this file to use in Google Earth, highlight the file by moving your cursor over the file. Click the **Share** button that appears on the right side. Click **Show link** and copy the URL that appears in the text box. Open Google Earth and paste this URL into the Image URL text field. **In the pasted Image URL, change "www" to "dl" within Google Earth (or within web browsers) if necessary**.

Please note that Dropbox continually updates its website and services, which might require minor modifications to these instructions and/or use of alternative web servers. For example, Dropbox recently has changed its user interface to improve the user experience, and it has discontinued serving html files in October 2016 (image files, for now, remain unaffected). If you use one of the many other services, you will likely go through similar steps of setting up an account, uploading files, and then obtaining the URL reference address for those files.

Please also recall that you can reference images already on the web by **clicking RMB on the image** and selecting **Copy Image Address/Location**. You should always first ask permission to reference someone else's images and realize that they may delete, move, or rename the images at any time, thus breaking your URL reference link to its location.

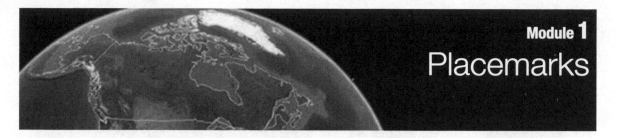

Placemarks

In this module, you will learn how to create and manipulate placemarks within Google Earth. **Placemarks** provide a means of saving a location with a specific view, zoom, and perspective, relaying textual and graphic content about the location, and much more!

Project 1.1	creating placemarks

1. Type <u>Mount Saint Helens</u> in the text field of the **Search** panel. Press **Enter/Return** or click the **Search** button to fly to this area. Placemarks available in Google Earth appear in the area

   option to save the search results to the My Places folder (**folder with arrow**), save the search results to the clipboard as a KML file (**clipboard**), print the search results (**printer**), or erase the search results (**blue X**). *Note: You can show your search history by clicking the **History** link. Similarly, you can get directions between two locations using the **Get Directions** link.*

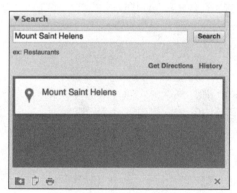

Figure 1.1: Search panel for
Mount Saint Helens.

2. Click on the **Add Placemark** icon ⟨☆⁺⟩ on the toolbar (or **Add > Placemark** from the menu). A new placemark appears in the Google Earth viewer with a yellow box around the icon.

3. In the **New Placemark** dialog, type <u>Mt. St. Helens</u> in the **Name** text field.

4. You can move the new placemark icon in the Google Earth viewer by moving your cursor over the blinking yellow box (the cursor should change to a hand with a pointing finger). Click and drag the icon to its proper location. For this project, move the placemark just to the NNE of the volcano to a position on Spirit Lake. *Note that it is not uncommon for the **New Placemark** dialog to obscure the new placemark icon; if it does, click on the dialog's title bar and drag it to one side.*

5. When you're finished, click **OK**. That's it! You've made a placemark! You can now visit somewhere else in Google Earth and instantly return to this location by simply locating this placemark in your **Places** panel and double-clicking it.

Where is my placemark stored?

Once you have clicked OK in the New Placemark dialog, the new placemark is typically added somewhere in the Places panel. To direct the placemark to be added to a specific folder, select the folder before adding the placemark or RMB on the folder and select Add > Placemark from the menu that appears. To copy a placemark, select the object to be copied, click RMB, and select Copy from the menu that appears. Then, select a folder, click RMB, and select Paste from the menu that appears. To store your placemark permanently, you should make sure that it is saved somewhere in your My Places folder.

Would you like to make this placemark more informative? We're going to learn how to do that in the next project on editing placemarks.

Why is text entered into the Name text field sometimes not kept?

This happens not only for the Name text field, but also occasionally for other text fields as well (e.g., Links to images, movies, etc.). The easiest workaround for this issue is to either tab out of the text field or simply click in another element of the dialog box (e.g., the Description text field). Your entries should then be retained.

Project 1.2   editing placemarks

1. Click RMB on the "Mt. St. Helens" placemark that we created in Project 1.1 (either in the **Places** panel or in the Google Earth viewer). A context menu appears with several operations that you can perform on this placemark. For this project, select **Get Info** (Mac) or **Properties** (PC). The **Edit Placemark** dialog appears (Figure 1.2).

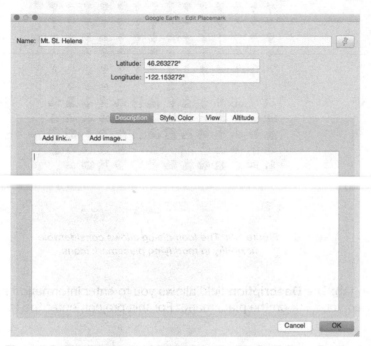

Figure 1.2: The Edit Placemark dialog for the Mt. St. Helens placemark.

2. You can edit the following features of your placemark:

- Name: Change the text in the **Name** text field to change the name of the placemark.

- Latitude & Longitude: These fields show the exact location of the placemark and will display the coordinates in the format specified in the **Preferences** (Mac) or **Options** (PC). In Figure 1.2, latitude and longitude are shown in decimal degrees.

  These numbers can be changed manually by entering values in the text fields, or they will automatically change as you move the placemark. *(Note: You must be in Edit mode, with a yellow box around the placemark, to move the placemark.)*

  What do positive and negative latitudes and longitudes mean?

  Positive numbers indicate latitudes <u>north</u> of the equator and longitudes <u>east</u> of the prime meridian. Conversely, negative numbers indicate latitudes <u>south</u> of the equator and longitudes <u>west</u> of the prime meridian.

- Icon: Click on the icon to the right of the **Name** text field to open the **Icon** dialog (Figure 1.3). Here you can select a different icon; use a custom icon; change the color, scale, or opacity of an icon; remove an icon; etc. Once you've made changes, click **OK**.

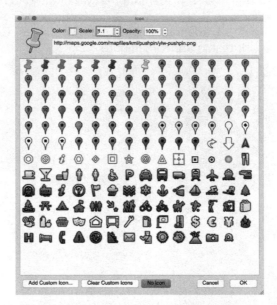

Figure 1.3: The Icon dialog allows considerable flexibility in modifying placemark icons.

Description tab: The **Description** field allows you to enter information that you want to pop up when you click on the placemark. For this project, enter the text in Figure 1.4.

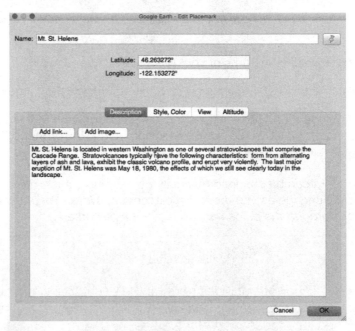

Figure 1.4: The Description field allows you to provide the user with information about your placemark.

118

- <u>Style, Color tab</u>: The **Style, Color** tab allows you to customize the color, size, and opacity of your placemark icon and its label. Please feel free to experiment with these settings on your own.

- <u>View tab</u>: The **View** settings allow you to customize the view that you see when you fly into a placemark. View information (latitude, longitude, range, heading, and tilt) may be entered manually, or you can set the view using the Google Earth viewer navigation controls and then click **Snapshot current view** to retain the zoom, perspective, and position. Zoom into the Spirit Lake area in the Google Earth viewer and tilt your perspective to the S such that you can see the topographic effects of the volcano (e.g., Figure 1.5) and then click **Snapshot current view** to change the settings (Figure 1.6).

*Figure 1.5: A perspective view to the south of the
Mount Saint Helens volcanic cone.*

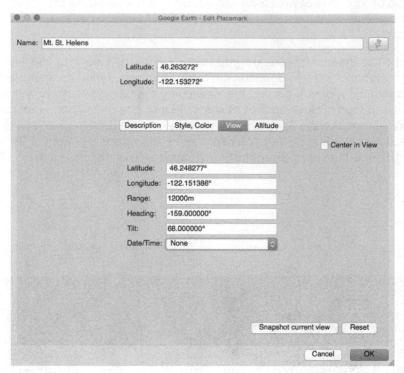

*Figure 1.6: Settings in the View tab created by clicking
Snapshot current view for the view in Figure 1.5.*

- Altitude tab: The **Altitude** tab specifies where the placemark icon is located relative to the map. Unless you want your icon floating above the Earth's surface, leave the altitude at 0 m. Please feel free to experiment with this setting on your own.

- Add link... button: The **Add link...** button inserts the formatted code at the cursor's location in the **Description field** to create a clickable URL within your placemark (see Project 1.5).

- Add image... button: The **Add image...** button writes the formatted code at the cursor's location in the **Description field** for embedding an image within your placemark (see Project 1.4).

3. Click **OK** to apply your changes and to close the **Edit Placemark** dialog.

4. Now zoom out until you can see the entire planet, and then double-click on your "Mt. St. Helens" placemark. You should now fly in to your location and stop at the perspective and zoom level that you set. In addition, the descriptive text that you entered for the placemark now appears in a pop-up balloon.

   Congratulations! You've just learned how to turn an ordinary placemark in Google Earth into an informative kiosk! Now, how about adding some pizazz to your placemark text by formatting it?

120

| Project 1.3 | basic formatting of placemark text |

Placemark text can be formatted using HTML tags, the same code that is used to format webpages.

1. Click RMB on the "Mt. St. Helens" placemark for Project 1.3, and select **Get Info** (Mac) or **Properties** (PC) from the context menu. The **Edit Placemark** dialog appears.

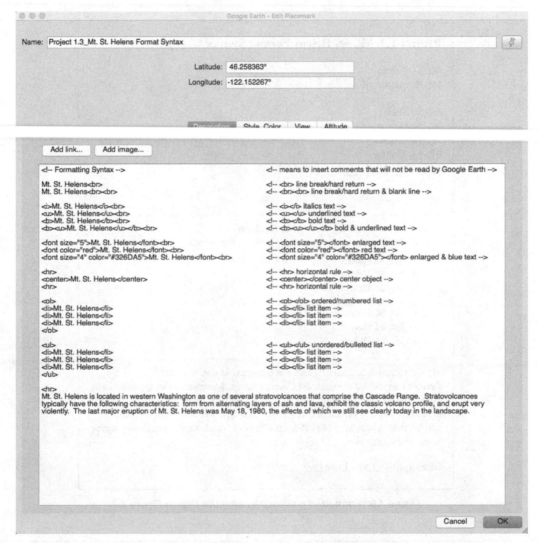

Figure 1.7: Basic HTML tags used to format placemark text.
Note that you can use some basic color names in place of the hexadecimal values.

2. Figure 1.7 shows the **Description** field with some basic formatting tags, and Figure 1.8 shows the resulting output.

121

What are formatting tags?

Formatting tags open and close around the text on which they act (and if you open something, you generally must close it). When several tags are nested (i.e., act on the same text), they must be closed in the reverse order in which they were opened. Common font colors can be specified by name, whereas other colors can be specified in hexadecimal format.

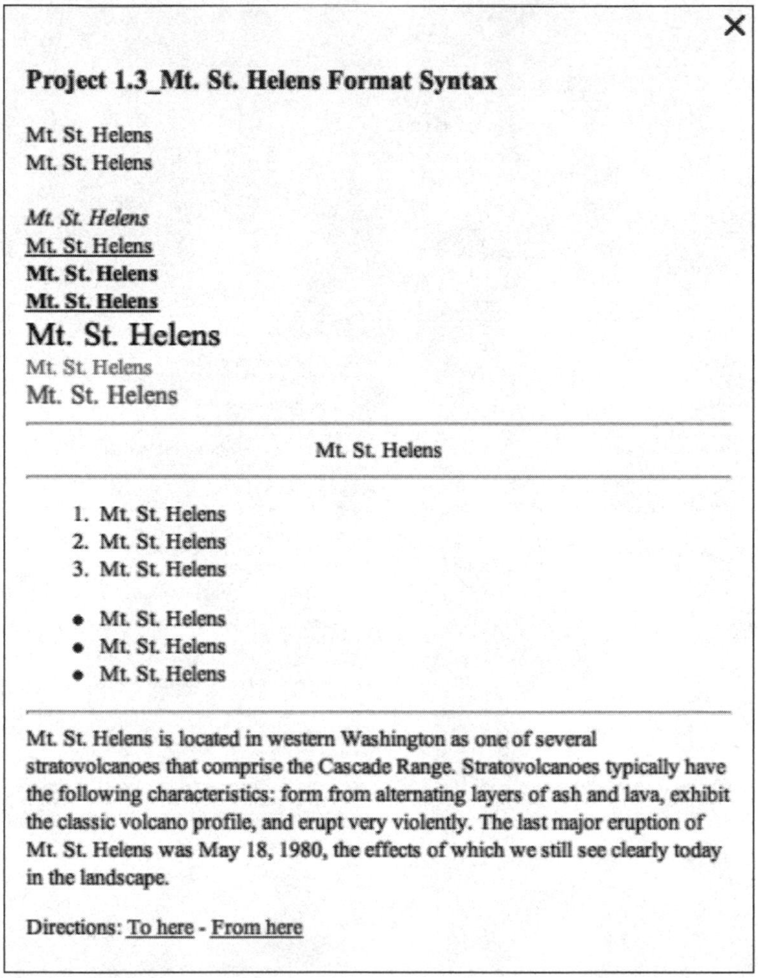

×

Project 1.3_Mt. St. Helens Format Syntax

Mt. St. Helens
Mt. St. Helens

Mt. St. Helens
Mt. St. Helens
Mt. St. Helens
Mt. St. Helens
Mt. St. Helens
Mt. St. Helens
Mt. St. Helens

Mt. St. Helens

---

1. Mt. St. Helens
2. Mt. St. Helens
3. Mt. St. Helens

- Mt. St. Helens
- Mt. St. Helens
- Mt. St. Helens

Mt. St. Helens is located in western Washington as one of several stratovolcanoes that comprise the Cascade Range. Stratovolcanoes typically have the following characteristics: form from alternating layers of ash and lava, exhibit the classic volcano profile, and erupt very violently. The last major eruption of Mt. St. Helens was May 18, 1980, the effects of which we still see clearly today in the landscape.

Directions: To here - From here

Figure 1.8: Formatted placemark using the tags in Figure 1.7.

How can I determine hexadecimal color values?
There are many free resources (e.g., online hexadecimal color charts, freeware [PC], & Digital Color Meter [Mac]) as well as commercial software (e.g., Adobe Photoshop) available.

122

3. Let's edit the "Mt. St. Helens" placemark again, and this time include the following in the **Description** text field:

```
<hr>
<center>
<font size="5" color="#4E7646">Mt. St. Helens</font><br>
<i>Cascade Range, western Washington</i>
</center>
<hr>
<b>Mt. St. Helens</b> is located in western Washington as one of several <i>stratovolcanoes</i>
that comprise the Cascade Range.  <i><u>Stratovolcanoes</u></i> typically have the following
characteristics:
<ol>
<li>form from alternating layers of ash and lava,</li>
<li>exhibit the classic volcano profile, and </li>
<li>erupt very violently.</li>
</ol>
<br>
The last major eruption of Mt. St. Helens was <b>May 18, 1980</b>, the effects of which we still see
clearly today in the landscape.
```

4. Figure 1.9 shows the resulting formatted placemark from the text in step 3.

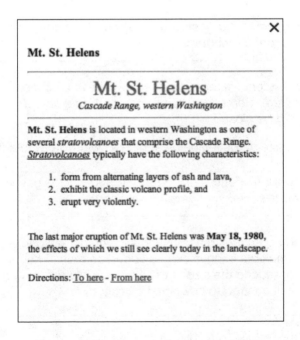

Figure 1.9: Formatted placemark using the tags in Step 3.

5. Great! You now have a basic understanding of how to format placemark text...but there is still more that you can do with placemarks! Before we go on, now would be a good time to practice some of the concepts that you've learned.

Project 1.4   inserting images into placemarks

Images within placemarks are an important way to convey additional information about a location, whether it be a graph of data, a photo, or a map. To insert an image into a placemark, you must first have its Internet URL address (that is, the image must be located on a web server). For most images on the Web, you can simply click RMB on the image in a web browser and copy its URL *(just realize that the availability of that image depends entirely on that website and web server).*

1. In a web browser, navigate to the Mount Saint Helens page on the USGS/Cascades Volcano Observatory website (http://volcanoes.usgs.gov/volcanoes/st_helens/). Click RMB on the photo of Mount Saint Helens on the right side of the page and copy the URL address for that image (http://volcanoes.usgs.gov/vsc/images/st_helens/Mount-st-Helens-home.jpg).

2. Click RMB on the "Mount Saint Helens" placemark that we edited in Project 1.3, and select **Get Info** (Mac) or **Properties** (PC) from the context menu. The **Edit Placemark** dialog appears.

3. In the **Description** text field, add " " tags, and place your cursor between the tags. Click the **Add image...** button, paste the URL text for your image in the URL text box that appears, and click OK to the right of the **Image URL** text box. Google Earth will insert the properly formatted <img src=" "> tags into your **Description** text field to create a centered image in your placemark (Figure 1.10). To change the image, simply change the URL in src="URL".

> * <img src="http://volcanoes.usgs.gov/vsc/images/st_helens/Mount-st-Helens-home.jpg"/> *

Note: If you omit the <center> tags, the image will be left-justified. Also, note that the closing tag </img> was omitted. Some tags that perform a single operation (e.g., just loading an image) do not need to be closed.

4. You can change the size of the image by adding a "width" attribute in pixels.

> * <img width="400" src="http://volcanoes.usgs.gov/vsc/images/st_helens/Mount-st-Helens-home.jpg"/> *

Congratulations! Inserting images really just entails adding this one line of code to your **Description** text field (multiple images can be added as well). The real work with images comes up front when you must reduce the size of the images while retaining reasonable quality such that the images are small enough to download quickly.

What if my image is a broken link?
There are many reasons that this might happen. The three most common are:
(1) you linked to an image that is on your computer and not on a web server,
(2) the name contains an unusual character/symbol (e.g., a space, a symbol other than "_", etc.), and (3) the URL link to the image is too long. Try to use short, descriptive names that do not contain spaces or special symbols.

Can I use images from my hard disk?

Yes...but you will be able to see the image only from that computer. Consider storing your images on a web server like Dropbox (see Before You Begin at the start of Section 3) if you want others to have access to your images.

The easiest way to find the **Image URL** on your hard disk for the **Add image...** dialog is to:
- locate the picture on your hard disk using a web browser (**File > Open**),
- copy the path from the web browser URL text box, and
- paste the path into Google Earth's **Image URL** text box.

Note: Google Earth will add "file:///" to the <img> tag, so you may need to experiment with the exact path you paste depending on your OS and browser.

Figure 1.10: Formatted placemark with a centered image.

Project 1.5   inserting URL links into placemarks

Sometimes you want to link to additional Internet content outside of Google Earth. In such instances, it would be nice to have a clickable link that opens a web browser to the appropriate page. For example, we would be remiss in not referencing the image that we obtained from the USGS/Cascades Volcano Observatory website. Here's how we do that.

1. In a web browser, navigate to the Mount Saint Helens page on the USGS/Cascades Volcano Observatory website and locate the photo of Mt. Saint Helens that we used. Click RMB on the photo and copy the URL address for that image (http://volcanoes.usgs.gov/vsc/images/ st_helens/Mount-st-Helens-home.jpg).

2. Click RMB on the "Mt. Saint Helens" placemark that we edited in Project 1.4, and select **Get Info** (Mac) or **Properties** (PC) from the menu. The **Edit Placemark** dialog appears.

3. In the **Description** text field after the </center> closing tag, add "*
<center> Image from USGS/Cascades Volcano Observatory website.</center>*". Select/highlight "USGS/Cascades Volcano Observatory", click the **Add link...** button, paste the URL location for your image in the URL text box that appears, and click OK next to the **Link URL** text box. Google Earth will insert the properly formatted <a href=" "> tags into your **Description** text field to create a clickable URL in your placemark (Figure 1.11). *Note: If you don't select text to become the link, Google Earth will insert the actual URL as the clickable text for your link.*

*
<center>Image from <a href="http://volcanoes.usgs.gov/vsc/images/ st_helens/Mount-st-Helens-home.jpg">USGS/Cascades Volcano Observatory</a> website.</center>*

Figure 1.11: Formatted placemark with a centered image & URL link.

Project 1.6   inserting YouTube video into placemarks

Video is yet another powerful means of conveying information about a place. Google Earth can use "embed" code within a placemark to link to web videos that offer that option. If there is not a direct embed link and if the web video is based on Adobe Flash, you might be lucky enough to copy the Flash embed code directly by clicking RMB on the web video and pasting it into a Google Earth placemark.

For this project, we are going to provide instructions on using web videos from the popular **YouTube** website. It is slightly more involved than what you learned about inserting images into placemarks in Project 1.4 because you have to obtain the embed code; however, it still is fairly straightforward.

> *WARNING: Due to compatibility issues between recent versions of Adobe Flash Player and Google Earth for the Mac, YouTube videos (Flash or HTML 5) will not display properly within placemarks on the Mac. Specifically, YouTube videos commonly either appear as white or black rectangles (clicking on the latter often crashes Google Earth on the Mac). There are two workarounds:*

> *1) create a URL link in the placemark to the video in order to view it in a web browser (not a bad idea even for PC users).*

> *2) temporarily downgrade your version of Adobe Flash Player to version 10.3.183.90 (circa 2013, download uninstaller/installer from https://helpx.adobe.com/flash-player/ kb/archived-flash-player- versions.html#Flash%20Player %20archives).*

1. In a web browser, go directly to a USGS video about Mt. St. Helens by typing the following URL (http://youtu.be/ sC9JnuDuBsU).

2. Click on the **Share** button beneath the video. An information frame expands. Click the **Embed** button and copy the highlighted code that appears in the expanded frame.

> Can I embed my own movies into Google Earth placemarks?
>
> Yes...however, it is a non-trivial process beyond the scope of this workbook. The most straightforward approach is to upload your video to YouTube and then use the technique described in Project 1.6.

3. In Google Earth, click RMB on the "Mt. Saint Helens" placemark that we edited in Project 1.3 (or Projects 1.4–1.5, although those will have the image that was inserted into the placemark) and select **Get Info** (Mac) or **Properties** (PC) from the context menu. The **Edit Placemark** dialog appears.

4. In the **Description** text field at the bottom of your description beneath the horizontal rule, add <center></center> tags and paste the YouTube embed code between them. Now enter a <br> tag after the embed code. Type Mount St. Helens: A Catalyst for Change | USGS YouTube Channel and create a URL link to the YouTube site by highlighting "*Mount St. Helens: A Catalyst for Change*" and adding the URL link from the Share button on the YouTube site (this not only allows Mac users to view the movie in a web browser, but it also references the video source). Give the video a few seconds to load as it is more than 6 minutes long. *(Note: If you omit the <center> tags, the video will be left-justified. Make sure that the pasted embed code has either "http" or "https" in the URL.)*

127

```
<center>
<iframe width="420" height="345" src="https://www.youtube.com/embed/
sC9JnuDuBsU" frameborder="0" allowfullscreen></iframe>
<br>
<a href="https://www.youtube.com/embed/sC9JnuDuBsU">Mount St. Helens: A
Catalyst for Change</a> | USGS YouTube Channel
</center>
<br><br>
```

5. Congratulations! You now have an enormous source of YouTube video material at your disposal to add to your placemarks (see Figure 1.12)!

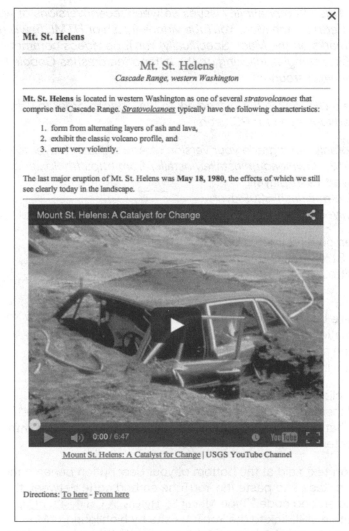

Figure 1.12: Formatted placemark with a centered YouTube movie.

Paths & Polygons

This chapter will introduce you to how you can use paths and polygons in Google Earth. **Paths** are lines or multi-segmented lines consisting of a sequence of individual points. They can be used to trace routes, to define the track for flyover tours, to outline interesting features, to measure the distance between points, etc. **Polygons** can be viewed as closed paths that define areas. They can be used to highlight specific regions (with or without transparency), and in Google Earth Pro, to measure areas (among other things).

Project 2.1	creating paths

1. For this project, we're going to first fly to Pompei, Italy. Type <u>Pompei, Italy</u> in the text field in the **Search** panel, and click the **Search** button to fly there.

2. Add a placemark named "Pompeii" and move it about 1000 m west of where Google marked the modern city of Pompei, Italy (yes, they are spelled differently!). Set an oblique perspective to the NW with an eye altitude of 775 m (Figure 2.1). This new placemark, as most archaeology buffs know, marks the ancient Roman city of Pompeii, which was partially buried in 79 C.E. by a pyroclastic eruption of the nearby volcano Mt. Vesuvius. A great deal of the city has now been excavated to provide a wonderful glimpse into everyday Roman culture and society.

Figure 2.1: Perspective view of the ancient Roman city of Pompeii with Mt. Vesuvius in the background.

3. Navigate to a view where you are looking vertically down on Mt. Vesuvius and Pompeii. We are going to create a path from the volcano to the city of Pompeii.

4. To create a new path, click the **Add Path** icon or click RMB in the **Places** panel and choose **Add > Path**. The **New Path** dialog appears, and your cursor changes to a crosshair in order to digitize points on the path. Name the path "Pyroclastic Flow".

5. Begin by clicking a point within Mt. Vesuvius's crater (you may have to move the **New Path** dialog). Then click a few more points to create a white path across the landscape from Mt. Vesuvius to Pompeii (Figure 2.2). Click **OK** to finish constructing the path.

Figure 2.2: Vertical view of the path constructed from Mt. Vesuvius to the ancient city of Pompeii.

Very good! In Module 4, we'll learn how to automatically fly a tour along this path to actually simulate the pyroclastic flow that destroyed Pompeii.

Project 2.2 — editing paths

1. Click RMB on the "Pyroclastic Flow" path that we created in Project 2.1 (either in the **Places** panel or in the Google Earth viewer). A context menu appears with several operations that we can perform on this path. For this project, select **Get Info** (Mac) or **Properties** (PC). The **Edit Path** dialog appears (Figure 2.3).

Figure 2.3: The Edit Path dialog for the Pyroclastic Flow path.

2. You can edit the following features of your path (many of them are similar to editing a placemark):

- Name: Change the text in the **Name** text field to change the name of the path.

- Description tab: The **Description** field allows you to enter information that you want to pop up when you click on the path. For this project, we'll not include any text here.

- Style, Color tab: The **Style, Color** tab allows you to customize the color, width, and opacity of the path line. For this project, change the color of the line to blue and increase its width to 2.0.

- View tab: The **View** tab really isn't all that beneficial for editing paths. The **Snapshot current view** function controls the view when you double-click the path.

- Altitude tab: The **Altitude** tab specifies where the path line is located relative to various surfaces associated with the map. For this project, select **Relative to ground** from the drop-down menu and enter "500" into the **Altitude** text field. Next, check the **Extend path to ground** checkbox to create a skirt beneath the line that extends to the ground (when this feature is active, you can change its style and color in the **Style, Color** tab).

- Measurements tab: The **Measurements** tab provides the length of the path in the specified units.

- Add link... button: The **Add link...** button inserts the formatted code at the cursor's location in the **Description** field to create a clickable URL within your path balloon (see Project 1.5).

- Add image... button: The **Add image...** button writes the formatted code at the cursor's location in the **Description** field for embedding an image within your path balloon (see Project 1.4).

- Last, and perhaps most important, the individual path points can be edited (you may need to move the **Edit Path** dialog out of your way). That is, the points that you digitized to create the path become highlighted in red when the **Edit Path** dialog is active. If you pass the cursor over any red point, it will turn green, and the cursor will turn into a hand with a finger. You then can adjust that point on the path.

3. Click **OK** to apply your changes and to close the **Edit Path** dialog. Figure 2.4 shows your results.

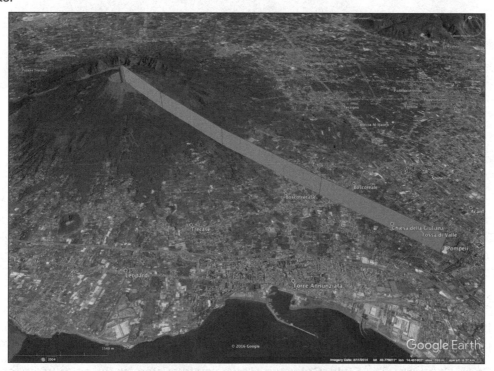

Figure 2.4: The Pyroclastic Flow path with some of its characteristics modified using the Edit Path dialog.

What is a Smoot?
Oliver Smoot '62 was the shortest pledge in the Lambda Chi Alpha fraternity at MIT in Cambridge, MA. In 1958, his fraternity brothers decided to use his 5-foot 7-inch (~1.70 m) frame as a measuring stick for the Harvard Bridge (which was 364.4 Smoots long). This "unit" of measurement has endured and has been included in Google Earth (http://en.wikipedia.org/wiki/Smoot and references therein).

Project 2.3   measuring lines & paths

1. For this project, we're going to first fly to the South Rim of the Grand Canyon. Type the coordinates <u>36.057015, -112.138989</u> in the text field in the **Search** panel, and click the **Search** button to fly there.

2. Click on the **Ruler** tool 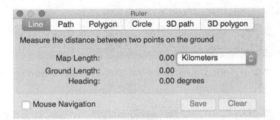 and select either **Line** or **Path.** (*Note: The other tabs are unique to Google Earth Pro. The Polygon tab will be explored later in this module.*)

   - The **Line** tab allows a line to be drawn between two endpoints and displays the **length** in the units in the drop-down menu and the **heading** in degrees (Figure 2.5). <u>**Map Length** provides the horizontal distance between the endpoints, whereas **Ground Length** crudely approximates the ground distance by taking into consideration the elevation and map distance of the endpoints.</u> *Step 4 (**Show Elevation Profile**) provides a more accurate measurement of the true ground distance.*

   - The **Path** tab permits you to draw a path composed of multiple points and to show the **cumulative length** in the units specified in the drop-down menu.

   - The Line or Path can be removed by clicking the **Clear** button or can be added as a new path using the **Save** button (in which case the **New Path** dialog appears).

Figure 2.5: Ruler dialog with the Line tab selected.

3. Use the **Path** tab in the **Ruler** to create, measure, and save a path between visitor centers on the North and South Rims (Figure 2.6; Map Length = 17.30 km and Ground Length = 17.31 km). You may want to turn on *More > Parks / Recreation Areas > US National Parks > Park Descriptions* in the **Layers** panel to see the visitor centers. You can also click the Mouse Navigation checkbox to use the navigation tools to zoom and pan to locate the endpoints.

4. Now click RMB on the "visitor centers" path that we just saved (either in the **Places** panel or in the Google Earth viewer). From the context menu that appears, you have two additional ways to obtain measurements for the path:

   - Select **Get Info** (Mac) or **Properties** (PC) from the context menu, and the **Edit Path** dialog appears (Figure 2.3). Select the **Measurements** tab to display the distance between the visitor centers (this is equivalent to the **Ground Length** displayed by the **Ruler** tool).

   - Select **Show Elevation Profile** from the context menu, and a new window that shows the elevation changes along the path appears at the bottom of your screen (Figure 2.7). As you move your mouse within the **Elevation Profile** window, a red arrow in the Google Earth window shows the corresponding point along your path. The display also shows the slope of the ground surface as well as the **actual terrain distance** along the path (*for a profile to show seafloor bathymetry for paths over oceans, select **Altitude > Clamp to*** *sea floor* in the *Edit Path* dialog box). Note that the terrain distance displayed in the **Elevation Profile** window is greater than the distances displayed in the **Edit Path** or **Ruler** dialogs because detailed topography is taken into account (20.9 km for the visitor centers' path). The vertical exaggeration and detail of the profile cannot be changed. Click the "X" on the right side of the **Elevation Profile** window to close it. *Note that the **Path** tab in the **Ruler** also has a **Show Elevation Profile** option.*

Figure 2.6: A path drawn using the Ruler tool and then saved as a path.

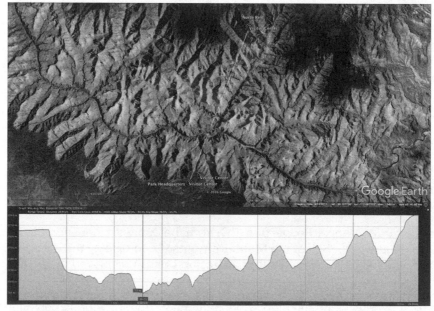

Figure 2.7: With Show Elevation Profile selected, users visualize the terrain and measure slope and actual terrain distances along a specified path.

134

Project 2.4   **creating polygons**

1. Type coordinates <u>35.027203, -111.022724</u> in the text field in the **Search** panel, and click the **Search** button to fly to Meteor Crater (aka Barringer Crater).

2. Navigate to a view where you are looking vertically down on the crater (Figure 2.8). We are going to create a polygon of the crater rim in order to estimate the area of the crater.

3. To create a new polygon, click the **Add Polygon** icon 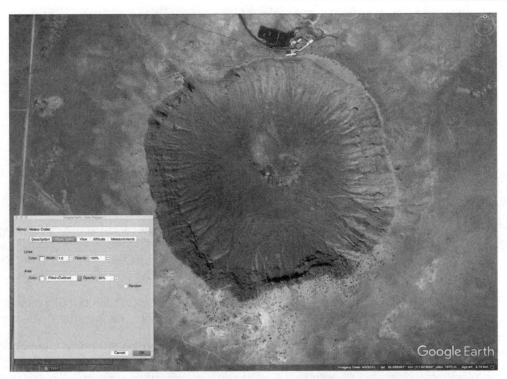 or right-click in the **Places** panel and choose **Add > Polygon**. The **New Polygon** dialog appears, and your cursor changes to a crosshair in order to digitize points on the polygon. Name the polygon "Meteor Crater". As you create the polygon, portions of the image may be obscured. If so, click on the **Style, Color** tab and set the **Area > Opacity** to **50%**.

4. Begin by clicking a point along the crater rim (you may have to move the **New Polygon** dialog). Then click additional points to create a white polygon that outlines the crater rim (Figure 2.8; the blue point is the last point added). Click **OK** when you finish constructing the polygon.

Figure 2.8: Vertical view of Meteor Crater during the initial stages of constructing a polygon to outline the crater rim.

Great! In the next project, we'll learn how to edit the polygon to change the points and to change the color of the polygon.

Project 2.5 editing polygons

1. Click RMB on the "Meteor Crater" polygon that we created in Project 2.4 (either in the **Places** panel or in the Google Earth viewer). A context menu appears with several operations that we can perform on this polygon. For this project, select **Get Info** (Mac) or **Properties** (PC). The **Edit Polygon** dialog appears (Figure 2.9).

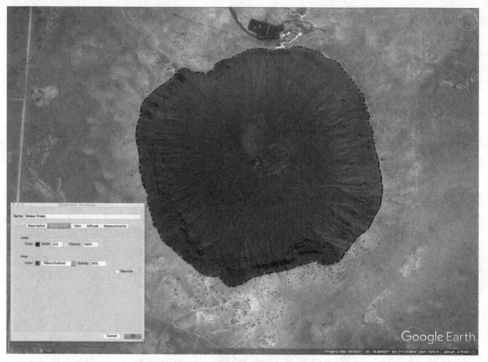

Figure 2.9: The Edit Polygon dialog for the Meteor Crater polygon.

2. You can edit the following features of your polygon (many of which are similar to editing a path):

   • Name: Change the text in the **Name** text field to change the name of the polygon.

   • Description tab: The **Description** field allows you to enter information that you want to pop up when you click on the polygon. For this project, we'll not include any text here.

   • Style, Color tab: The **Style, Color** tab allows you to customize the color, width, and opacity of the polygon outline and fill (area). For this project, change the color of the line to black and increase its width to **2.0**. Change the color of the area to blue and make the area have a **50% opacity**.

   • View tab: The **View** tab really isn't all that beneficial for editing polygons. The **Snapshot current view** function controls the view when you double-click the polygon.

   • Altitude tab: The **Altitude** tab specifies where the polygon is located relative to various surfaces associated with the map. For this project, leave it as **Clamped to ground**.

   • Measurements tab: The **Measurements** tab provides the perimeter and area of the polygon in the specified units (Google Earth Pro feature only).

- Add link... button: The **Add link...** button inserts the formatted code at the cursor's location in the **Description** field to create a clickable URL within your polygon balloon (see Project 1.5).

- Add image... button: The **Add image...** button writes the formatted code at the cursor's location in the **Description** field for embedding an image within your polygon balloon (see Project 1.4).

- Finally, just as for paths, the individual points that compose the polygon can be edited (you may need to move the **Edit Polygon** dialog out of your way). That is, the points that you digitized to create the polygon become highlighted in red when the **Edit Polygon** dialog is active (Figure 2.9). If you pass the cursor over any red point, it will turn green, and the cursor will turn into a hand with a finger. You then can adjust that point on the polygon outline. Additional points can be added by clicking on the line next to the active (blue) dot.

3. Click **OK** to apply your changes and to close the **Edit Polygon** dialog.

Project 2.6 | measuring polygons

1. For this project, we're going to assume that we didn't create a new polygon in Project 2.4, but that we still want to know the perimeter and area of the crater at Meteor Crater.

2. Click on the **Ruler** tool ▨ and select the **Polygon** tab (Google Earth Pro only). Your cursor changes to a crosshair in order to digitize points on the polygon. As you digitize points, the Ruler dialog displays the perimeter and the area in the units in the drop-down menus (Figure 2.10).

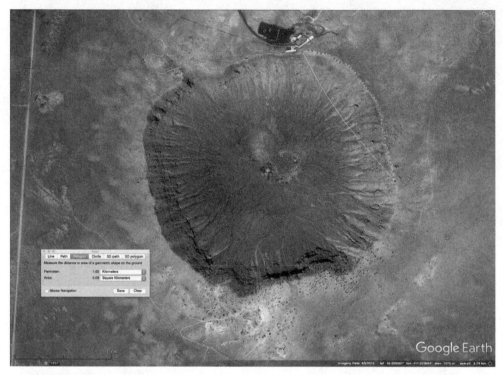

Figure 2.10: Vertical view of Meteor Crater during the initial stages of constructing a polygon to measure the perimeter and area of Meteor Crater using the Ruler tool.

3. The Polygon can either be removed by clicking the **Clear** button or added as a new polygon using the **Save** button (in which case the **New Polygon** dialog appears).

4. Recall that if you already have created a polygon, you can click RMB on the polygon (either in the **Places** panel or in the Google Earth viewer). From the context menu that appears, you can select **Get Info** (Mac) or **Properties** (PC) from the context menu, and the **Edit Polygon** dialog will appear. Select the **Measurements** tab to display the perimeter and area of this existing polygon.

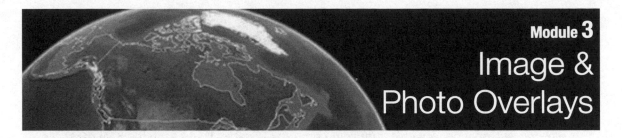

Image & Photo Overlays

Image and photo overlays are powerful features within Google Earth. **Image overlays** are graphics (e.g., maps, diagrams, etc.) that can be draped onto the landscape and georeferenced to their proper positions on the globe. **Photo overlays** are zoomable photos that can be embedded within the Google Earth viewer as virtual panoramas (e.g., rectangular "billboards", cylinders, spheres), which can be explored by the user.

Project 3.1	creating an image overlay

1. For this project, we're going to first fly to the Yucatán Peninsula. Type coordinates 20.188160, -88.859868 in the text field in the **Search** panel, and click the **Search** button to fly there. Stop your descent at an altitude of ~600,000 m (~370 mi).

2. In the **Layers** panel, click **Borders and Labels** to turn on the region outlines.

3. To create a new image overlay, click the **Add Image Overlay** icon 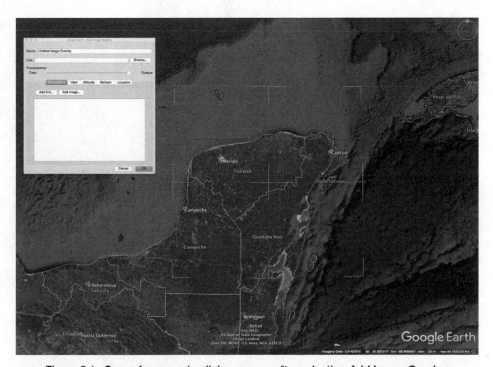 or click RMB in the **Places** panel and choose **Add > Image Overlay**. The **New Image Overlay** dialog and a green frame for your image appear (Figure 3.1).

Figure 3.1: Green frame and a dialog appear after selecting Add Image Overlay.

4. Images must be linked to the green frame in order to be displayed (see the **Link** text field in Figure 3.1). Such links can be created one of two ways:

   a. **From your local hard disk**: Click the **Browse** button, and a standard **Open File** dialog for your computer system will appear, allowing you to link to an image (e.g., JPG, GIF, TIFF, PNG, etc.) on your computer. *The only issue with this approach is that the image overlay will only be viewable from your computer.* That is, if you save the KML/KMZ file and move it to another computer, you'll only see a box with a large "X" instead of your image overlay (you'll also see the "X" if you move your image file and don't update the Google Earth link).

   b. **From the Web**: Type or paste a URL for an image that is on a web server in the **Link** text field.   For example, you can click RMB on most images on webpages, select **Copy Image Address/Location**, and paste it in the **Link** text field (of course, you should first request permission to use the image). Alternatively, for your personal images, you will need to load them to a web server where they can be referenced by a URL (see **Before You Begin** at the beginning of **Section 3** to learn about how to use Dropbox, or contact your technical support person about setting up a local web server).

5. In the **New Image Overlay** dialog, change the **Name** to "Satellite Overlay" and insert "http://media.wwnorton.com/college/geo/geotours2/ChicxulubSatImage.jpg". Click **OK**.

That is all there is to it! You'll notice, however, that the image overlay is not scaled and aligned to exactly match the yellow country borders. In Project 3.2, we'll learn how to edit the image overlay to georeference it with the Google Earth virtual globe.

Project 3.2   editing an image overlay

1. Click RMB on the "Satellite Overlay" overlay that we created in Project 3.1 (in the **Places** panel). For this project, select **Get Info** (Mac) or **Properties** (PC). The **Edit Image Overlay** dialog appears (Figure 3.2).

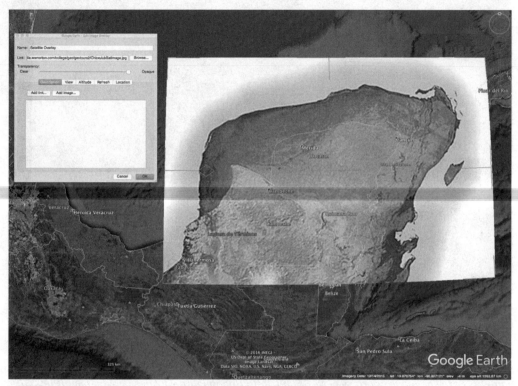

Figure 3.2: The Edit Image Overlay dialog with the green frame containing the non-georeferenced image overlay. Satellite image courtesy of NASA/JPL (http://photojournal.jpl.nasa.gov/catalog/PIA03379).

2. The editing objective is to manually adjust the image overlay to where it matches (as best you can) the area that it represents. In this example, we'll georeference the image by aligning the shoreline of the image with the yellow border outlining the Yucatán Peninsula. You can facilitate this process by moving the **Transparency** slider in the **Edit Image Overlay** dialog to a semi-transparent setting in order to better see how to align the image (you may want to move the **Edit Image Overlay** dialog out of the way).

3. To align the image, you will manipulate the green frame. To do so, move the cursor over the various frame features, and the cursor will change from the panning cursor (an open hand) to the editing cursor (a hand with a pointing finger) when it is over an editable part of the frame:

   • **Corners**: Changing the position of each corner scales the frame by moving the two adjacent sides. If you hold down the Shift key, the entire frame is scaled proportionally.

   • **Sides**: Changing the position of a side scales the frame by moving only that side. If you hold down the Shift key, the entire frame is scaled proportionally.

   • **Center Cross**: Changing the position of the center cross moves/translates the entire frame.

   • **Diamond** (left side of frame): Clicking and dragging within the diamond rotates the frame.

141

4. Figure 3.3 shows a georeferenced image overlay that has been adjusted "by eyeball" to obtain a reasonable fit (yours should look something like this). Please note that some images (e.g., maps, aerial photos, etc.) may not fit perfectly because of their projection or because of drafting errors in generating the map. Before you select **OK** to accept your adjustments, you probably will want to move the **Transparency** slider back to **Opaque**.

Figure 3.3: A semi-transparent overlay adjusted to be georeferenced to the Google Earth virtual globe country borders. Satellite image courtesy of NASA/JPL (http://photojournal.jpl.nasa.gov/catalog/PIA03379).

5. In the **Places** panel, select "Satellite Overlay," and click the **Adjust Opacity** icon at the bottom of the **Places** panel. Moving the slider that appears to the left makes the image overlay more transparent. You can adjust your viewing position and the transparency to highlight features on the ground surface related to your image overlay.

Congratulations! You have just learned a very important process you can use to integrate all types of maps, diagrams, and photos with the Google Earth imagery—and see them in their proper spatial perspective. Project 3.3 will show you how to take this information and use it in a meaningful way. Before we conclude, however, please note that **Google Earth Pro can automatically import georeferenced GEOTIFF images and shapefiles using the File > Import... menu.**

Project 3.3 — measuring objects on an image overlay

1. Turn on the "Satellite Overlay" overlay that we georeferenced in Project 3.2 (click the checkbox in the **Places** panel). Note that the NW corner of the Yucatán Peninsula shows a semi-circular feature. This feature is a partial expression of the Chicxulub (CHICKS-oo-loob) Crater that formed when an asteroid hit this region approximately 65.5 million years ago, leading to the extinction of many species (most famously, the dinosaurs).

2. To further constrain the geometry of the crater, we'll create an image overlay that depicts the gravity signature of this area and the nearby offshore. Using the techniques described in Projects 3.1–3.2, create and georeference a gravity map of the region found at http://media.wwnorton.com/college/geo/geotours2/ChicxulubGravityMap.gif. The resulting image overlay is shown in Figure 3.4.

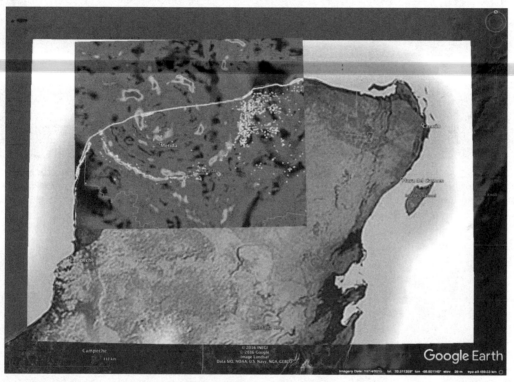

Figure 3.4: A gravity map image overlay of the Chicxulub Crater on top of the satellite image overlay of the area created and georeferenced in Projects 3.1–3.2. Satellite image courtesy of NASA/JPL (http://photojournal.jpl.nasa.gov/catalog/PIA03379). Gravity map reproduced with the permission of Natural Resources Canada, courtesy of the Geological Survey of Canada.

3. Using the bull's-eye pattern on the gravity map and the maximum extent of the crater as depicted on the satellite imagery overlay, you can now add a polygon (Project 2.4) to determine the crater's area (Figure 3.5). *If the polygon is not displayed on top of the overlays, edit the polygon by going to the Altitude tab, selecting Absolute, and entering a value of 500 m.*

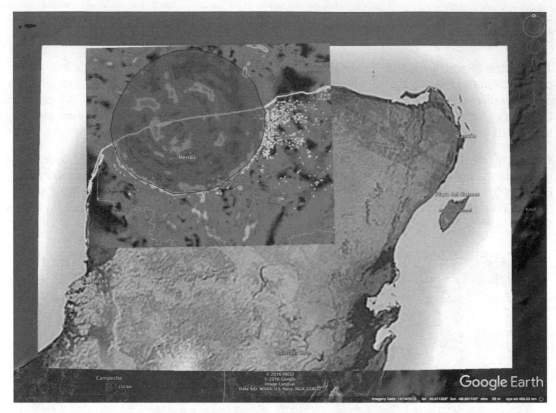

Figure 3.5: A semi-transparent polygon representing the Chicxulub Crater has a measured area of ~17,673 km² (compare this with the area of Meteor Crater in Project 2.5). Satellite image courtesy of NASA/ JPL (http://photojournal.jpl.nasa.gov/catalog/PIA03379). Gravity map reproduced with the permission of Natural Resources Canada, courtesy of the Geological Survey of Canada.

Project 3.4   creating a zoomable photo overlay

1. For this project, we're going to first fly to Yosemite National Park. Type the coordinates 37.746773, -119.591832 in the text field in the **Search** panel and click the **Search** button to fly there. Rotate your perspective until you are viewing Half Dome toward the NE, as shown in Figure 3.6.

***Figure 3.6: Oblique perspective view of Half Dome from the Yosemite Valley floor.
Vertical exaggeration of the terrain should be 1.***

Half Dome is an exfoliation dome composed of granite/granodiorite of the Sierra Nevada batholith. It is rounded because the igneous rock exfoliates in concentric layers much like peeling the layers of an onion. The name *Half Dome* originated because the flat NW face makes it appear as though half of the dome is missing. This face is one of the systematic fractures that exists in the area.

We're going to create a zoomable photo overlay to better visualize this amazing landform.

2. To create a new photo overlay, click RMB in the **Places** panel and choose **Add > Photo** (or select **Add > Photo** from the menu bar). The **New Photo Overlay** dialog appears (Figure 3.7).

3. Change the name to "Half Dome" and enter the URL of an image in the **Link** text field (http://media.wwnorton.com/college/geo/geotours2/HalfDome.jpg; Figure 3.7).

Figure 3.7: Photo overlay of Half Dome.

4. Click **OK** to accept the photo overlay.

5. If you double-click on the Half Dome photo overlay in the **Places** panel, you will zoom into the photo, and the controls in the upper right corner of the Google Earth viewer will change. You can zoom/unzoom using SCROLL, or you can use the +/- controls in the upper right corner. The thumbnail in the upper right corner will develop a white outline as you zoom in on the photo. You can click and drag the white box to move your viewing location, or you can use the panning cursor (open hand). Click the **Exit Photo** button to exit the Zoomable Photo Overlay mode and to return to the Google Earth viewer. See Figure 3.8 to view the zoomable photo overlay controls.

*Figure 3.8: Double-clicking a photo overlay in the Places panel
allows the user to zoom into the photo at high resolution.*

Project 3.5   editing a zoomable photo overlay

1. Click RMB on the "Half Dome" photo overlay that we created in Project 3.4 (in the **Places** panel). Select **Get Info** (Mac) or **Properties** (PC). The **Edit Photo Overlay** dialog appears (Figure 3.9).

Figure 3.9: Clicking RMB on a photo overlay in the Places panel allows the user to edit the photo overlay.

2. You can change the positioning and perspective of the photo overlay using the controls in the **Photo** tab (Figure 3.9). This tool is very useful when aligning the photo with the Google Earth imagery. In Figure 3.9, the photo overlay was made semi-transparent to help with the alignment. Because the photo was taken from the ground, **Altitude** of the overlay was set to 2 m. The **Heading** (left-right), **Tilt** (up-down), and **Roll** (rotation) controls help align the position of the view behind the image by changing the camera view, whereas the **Horizontal** and **Vertical** controls help scale the view to the image. The vertical exaggeration in Google Earth should be set to 1 in order to minimize distortion *[Google Earth > Preferences (Mac) or Tools > Options (PC)].* Although aligning photo overlays is a trial-and-error process that requires some patience (especially if you don't know the exact location where the photo was taken), it can be a very powerful tool for facilitating detailed analysis of a feature in Google Earth (e.g., creating virtual "outcrops" for more detailed study of an area).

Tours & Historical Imagery

Google Earth has several tools that allow you to take automated tours and to view historical imagery for an area. **Tours** allow users to fly over the landscape in a prescribed manner (e.g., using the mouse as a joystick, a folder of placemarks (including showing placemark balloons), tracing a path, etc.). **Historical imagery** provides a wealth of information about changes to the landscape over time. Users can access available images in the Google Earth viewer and create placemarks to automatically return to that image.

Project 4.1	creating tours of a folder of placemarks

Google Earth allows you to select a folder in your **Places** panel and take an automated flying tour from placemark to placemark.

1. You can click once on any folder in your **Places** panel that contains placemarks to select it. For this project, we're going to use the "Project 4.1_Placemarks from Chap 10" folder in the **4. Tours & Historical Animations** folder (see **Places** panel in Figure 4.1). This folder of placemarks was created from John Wesley Powell's book, *The Exploration of the Colorado River and Its Canyons* (1997, Penguin Books), which chronicles the 1869 journey of Powell and his men down the unexplored and unmapped Colorado River in three wooden boats. Powell's book was derived from his journal, in which he provided dated entries keyed to prominent features along the river.

Figure 4.1: Placemarks from John Wesley Powell's book, <u>The Exploration of the Colorado River and Its Canyons</u>, document part of his 1869 journey down the unexplored and uncharted Colorado River.

If you want to create your own folder of placemarks, click RMB in the **Places** panel and select **Add > Folder**. Name the folder, then select it with a single mouse click. Any new placemarks created will go into this selected folder as long as it is highlighted. Alternatively, you can either drag or copy existing placemarks to the folder as well.

2. After you've highlighted the folder of placemarks, click on the **folder icon with the small triangle** (Figure 4.1; the triangle looks like a DVD player's Play button) in the lower right corner of the **Places** panel.

3. The **Tour Control** (Figure 4.2) will appear in the lower left corner of the Google Earth viewer window. The tour will begin with the first placemark and then will move sequentially to other placemarks.

> **Can I control the tour view?**
>
> Absolutely! You can edit each placemark's view as described in Module 1 to set the perspective, altitude, etc. You also can control aspects of the tour by using Google Earth > Preferences (Mac)/Tools > Options (PC) and then changing settings in the first section of the Touring tab. One neat feature is that you can automatically show your placemark balloons during the tour and specify their duration.

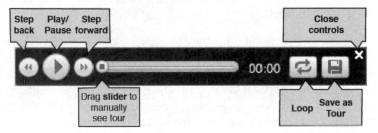

Figure 4.2: Tour Control.

4. The tour can be paused, looped continuously, or saved as a separate Google Earth tour in which you no longer need the placemarks (click the **Disk** icon in the **Tour Control** after playing the tour, see Project 4.2). If the tour isn't set to continuously loop, then it will stop and you can close the Tour Control by clicking the X. *Note: During the tour, the controls will become hidden so as to not obscure the view. To make them return, just move your mouse in the Google Earth viewer window.*

> **How can I make the placemarks invisible for tours?**
>
> It actually is fairly easy. Click RMB on the placemark in the Places panel and select Get Info (Mac) or Properties (PC). Click on the icon to change the icon to "No Icon" and name the placemark " " (an empty space). Alternatively, just play the tour and then save it using the Disk icon in the Tour Control, and the placemarks are no longer needed (see Project 4.2).

Project 4.2   creating tours along a specified path

Google Earth also offers the opportunity to fly tours along a prescribed path from the **Places** panel.

1. For this project, we're going to use the path that we created in Module 2 for the Roman city of Pompeii, Italy, which was partially buried by the eruption of Mt. Vesuvius. To fly to the region, double-click the "Pyroclastic Flow" path that you created. It is probably useful to check both the "Pyroclastic Flow" path and the "Pompeii" placemark so that they appear in the Google Earth viewer (Figure 4.3).

Figure 4.3: Vertical view of the path from Mt. Vesuvius to the ancient city of Pompeii.

2. We now want to fly along this path from the volcano to the city of Pompeii to simulate the 79 C.E. pyroclastic flow eruption of nearby Mt. Vesuvius that engulfed Pompeii. First, click the "Pyroclastic Flow" path in the **Places** panel once to highlight it.

3. The **Path Tour** icon will appear at the lower right corner of the **Places** panel. Click on the **Path Tour** icon to automatically fly along the path from the first to the last point.

151

4. After taking an initial tour, you probably will want to adjust the settings. To do so, select **Google Earth > Preferences** (Mac) or **Tools > Options** (PC) and select the **Touring** tab (Figure 4.4).

<u>What tour settings should I use? (in the Google Earth > Preferences (Mac) or Tools > Options (PC))?</u>

There is no hard-and-fast rule. The best recommendation is to simply fly the tour once to see how the angle, height, and speed of the camera appear. This also gives the Google Earth imagery an opportunity to load into memory, giving you a sharper image.

5. For this project, set the **Camera Tilt Angle** = 65, **Camera Range** = 100, and **Speed** = 900 (Figure 4.4).

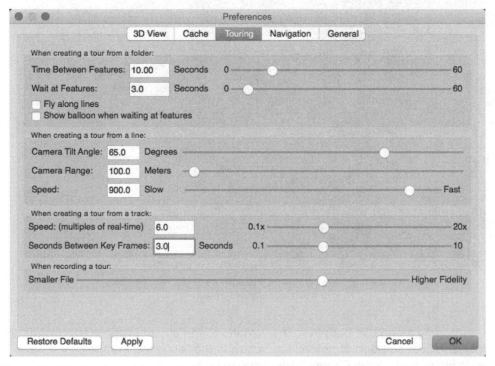

Figure 4.4: The Touring tab in the Preferences dialog. The middle section controls tours of paths.

6. Now play the tour again (but leave the "Pyroclastic Flow" path unchecked so that it is not visible in the Google Earth viewer). Note that the **Tour Control** appears (Figure 4.5).

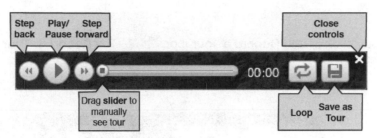

Figure 4.5: Tour Control.

7. After the tour has played, you can save it by clicking the **Disk** icon in the **Tour Control** tool. This allows you to delete the path and just retain the tour for subsequent use.

Excellent! You have just simulated the pyroclastic flow that destroyed Pompeii. It is interesting to note that the way the tour "erupts" from the volcano, how it bounces along the terrain toward Pompeii, and the speed at which the tour progresses is a reasonable facsimile of an actual pyroclastic flow.

Project 4.3   **creating tours from the Google Earth viewer**

Not only can you save tours of placemarks and paths, but you also can record tours of the Google Earth viewer.

1. To create a tour, click on the **Record a Tour** icon . The **Record a Tour** control appears in the lower left corner of the Google Earth viewer window (Figure 4.6).

Figure 4.6: Record Tour control.

2. To turn tour recording on, click the red **Record** button. Anything that occurs in the Google Earth viewer will be recorded. To record audio during the tour, click the blue **Record Audio** button (microphone). When you are finished recording, click the red **Record** button again. The **Tour Control** (Figure 4.7) appears, and your recorded tour will play. You can save the tour to your **My Places** folder in the **Places** panel by clicking the **Disk** icon in the **Tour Control**.

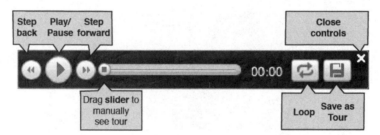

Figure 4.7: Tour Control.

That's it! You can now make recordings, save them to your **My Places** folder, and record audio narration to accompany the tour.

Project 4.4   viewing historical imagery I

1. In the text field in the **Search** panel, type <u>18.543157, -72.338851</u>, and click on the **Search** button to fly to the National Palace in Port-au-Prince, Haiti (use an eye altitude of approximately 430 m). This area was devastated by a 7.0 earthquake on January 12, 2010.

2. Click on the **Historical Imagery** icon to see pre-earthquake imagery. A **time span control** (Figure 4.8) will appear in the upper left corner of the Google Earth viewer.

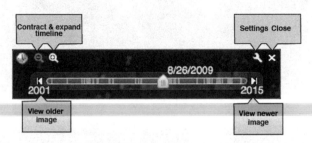

Figure 4.8: Time span control for historical imagery.

3. On the time span control, click the **button containing the left-pointing triangle (View Older Image** button) until the imagery date reads "8/26/2009". This is a pre-earthquake image of the area around the National Palace (Figure 4.9). *Please note that imagery dates in Google Earth are displayed according to user preference, so your dates may not exactly match the dates displayed in this workbook. Because imagery dates are stored in UTC (Coordinated Universal Time), the dates shown are more consistent if you click the **Wrench** icon (Figure 4.8) and set the date display to **UTC**.*

Figure 4.9: August 26, 2009, pre-earthquake image of the National Palace in Port-au-Prince, Haiti.

4. Now click the **button containing the right-pointing triangle (Step Forward** button) or drag the slider until the Imagery Date reads "1/17/2010". This shows a post-earthquake image of the area around the National Palace (Figure 4.10) a few days after the earthquake happened.

*Figure 4.10: January 17, 2010, post-earthquake image of the
National Palace in Port-au-Prince, Haiti.*

What happened to the time span control?

The time span control may disappear somewhat so it
does not obscure the image. To bring it back, pass the
mouse over the upper left corner where the control is
located.

That is all there is to accessing Google Earth's historical imagery. The time range and image
quality vary substantially from location to location. In the instance of the Haiti earthquake
tragedy, the January 17, 2010 image is extremely high resolution and was beneficial in helping
relief efforts in the region. In the next project, we'll investigate another example where historical
imagery tells an important story.

Project 4.5   viewing historical imagery II

1. In the text field in the **Search** panel, type -65.37, -60.95, and click on the **Search** button to fly to the Larsen B ice shelf in Antarctica. Make sure that your eye altitude is about 190 km. Turn off **View > Water Surface** (if necessary).

2. Click on the **Historical Imagery** icon and set the time span control (Figure 4.8) to "1/31/2002" *(please note that Google currently has some display issues on some dates and at higher elevations)*. Figure 4.11 shows the Larsen B ice shelf with some ponding of water on the surface.

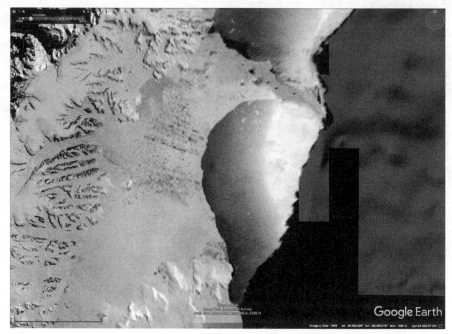

Figure 4.11: January 31, 2002 image of the Larsen B ice shelf in Antarctica.

3. Now click the **button containing the right-pointing triangle (Step Forward** button) or drag the slider until the Imagery Date reads "3/7/2002" (Figure 4.12) and then "2/23/2006" (Figure 4.13). This sequence of images from 2002 until 2006 captures the breakup of the Larsen B ice shelf.

> What happened to the ice shelf?
>
> Scientists believe that the polar ice caps and ice shelves are like canaries in coal mines—they are harbingers of the climatic effects of global warming.

Figure 4.12: March 7, 2002 image of the Larsen B ice shelf in Antarctica.

Figure 4.13: February 23, 2006 image of the Larsen B ice shelf in Antarctica.

Leftovers are those last remaining tidbits of "good stuff" that are just too good to throw out (so we grouped them together in this module)! Some of our leftovers are a bit more advanced and require editing KML files using a text editor, but hopefully they will help you make the transition into additional programming/exploration with KML files on your own.

Project 5.1	**importing GPS data into Google Earth**

1. GPS data can be imported into Google Earth in three basic ways:

    • **Importing a .gpx data file directly into Google Earth**.

    • **Importing data directly from a GPS device** (works for most Garmin and Magellan GPS units).

    • **Using a third-party application to create a KML file from a .gpx data file or from data imported directly from your GPS device** (e.g., http://www.gpsbabel.org/ or http://www.gpsvisualizer.com/). *Please note that instructions for using third-party applications are beyond the scope of this workbook (both sites provide excellent tutorials).*

2. Connecting your GPS unit to a computer.

    a. If you plan on downloading GPS data on a Windows computer, first install the USB driver that comes with your GPS unit. Garmin GPS users should be able to download this driver from the Garmin website: http://www8.garmin.com/support/download.jsp. For Magellan GPS users, the download support site is http://support.magellangps.com/support/index.php?_m=downloads&_a=view.

    b. With the GPS *off*, hook the GPS unit to the computer running Google Earth using the provided cable. Turn the GPS *on*.

3. Within Google Earth, select **Tools > GPS**, and the **GPS Import** dialog appears (Figure 5.1).

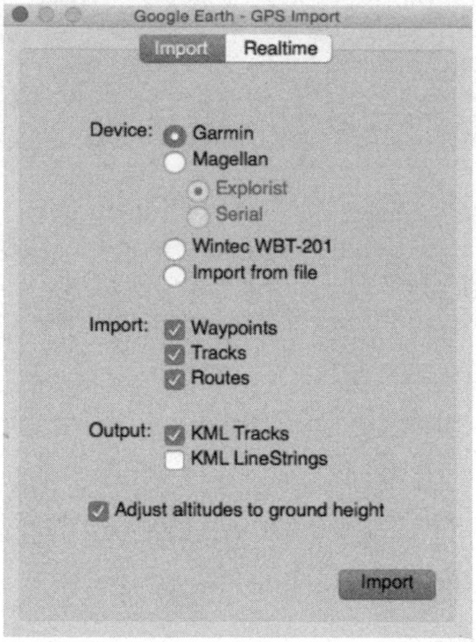

Figure 5.1: GPS Import dialog for downloading data directly from the GPS unit.

4. Select the appropriate **Device** (**Garmin**, **Magellan**, or **Import from File**).

5. Select the data types to **Import**:

    a. **Waypoints**—specific point locations recorded by the GPS user. Waypoints are represented in Google Earth as placemarks.

    b. **Tracks**—series of points automatically recorded by the GPS unit at a specified interval (sometimes referred to as breadcrumb trails). Tracks are represented in Google Earth as paths.

    c. **Routes**—series of points used by the GPS in routing (e.g., like using the "go-to" instructions). Routes are represented in Google Earth as paths.

6. In the **Output** section of the **GPS Import** dialog (Figures 5.1 and 5.2), users can choose:

    a. **KML Tracks**—Paths for imported **Tracks** and **Routes** will include a time element that can be animated as described in Projects 4.4 and 4.5.

    b. **KML LineStrings**—Paths for imported **Tracks** and **Routes** will not include a time element. Points composing the path will be displayed and will have a time element.

7. A checkbox allows you to **Adjust altitudes to ground height**, which automatically converts all imported data to lie on the ground surface in the Google Earth viewer.

162

8. When you click **Import**, a dialog appears detailing the data imported into Google Earth (if you are doing **Import from File**, you will first see a system dialog box that allows you to browse for the file). The data will be stored in your **Temporary Places** folder in the **Places** panel within a folder "garmin GPS device" or similar (Figure 5.2). These data must be moved to **My Places**, or they will be lost when you quit Google Earth.

Figure 5.2: Imported GPS data displayed in Google Earth viewer (and stored in the Temporary Places folder).

Project 5.2   **hiding description text in the Places panel**

1. Click RMB on the feature (either in the **Places** panel or in the Google Earth viewer). Select **Save Place As...**, and a standard **Save** dialog appears.

2. Select **KML** from the **Format** drop-down menu.

3. Open the KML file with a plain-text editor like TextEdit (Mac) or NotePad (PC) and edit the feature's KML code as follows (this example shows the syntax for a placemark, but will work for other features as well):

```
<Placemark>
      <Snippet maxLines="0"></Snippet>
   ...
   </Placemark>
```

Save the file with a .kml extension and reopen it in Google Earth (Figure 5.3; see bold text in example KML code provided after Project 5.4).

Project 5.3   **hiding directions & the default title in placemark balloons**

1. Click RMB on the placemark (either in the **Places** panel or in the Google Earth viewer). Select **Save Place As...**, and a standard **Save** dialog appears.

2. Select **KML** from the **Format** drop-down menu.

3. Open the KML file with a plain-text editor like TextEdit (Mac) or NotePad (PC) and edit the feature's KML code as follows: (*Note: This code must be inside all **Style** tags in the file.*)

```
<Style>
   ...
     </IconStyle>
     <BalloonStyle>
        <text>$[description]</text>
     </BalloonStyle>
   </Style>
```

4. Save the file with a .kml extension and reopen it in Google Earth (Figure 5.3; see bold text in example KML code provided after Project 5.4).

Project 5.4 — inserting comments in KML files or descriptions

1. Click RMB on the feature (either in the **Places** panel or in the Google Earth viewer). Select **Save Place As...**, and a standard **Save** dialog appears.

2. Select **KML** from the **Format** drop-down menu.

3. Open the KML file with a plain-text editor like TextEdit (Mac) or NotePad (PC) and edit the feature's KML code as follows *(note that it cannot be the first line in a KML file)*:

> **<!-- insert comment here -->**

Can comments be added to descriptions?

Absolutely! Just click RMB on the feature in the Places panel and select Get Info (Mac) or Properties (PC). In the Description tab, you can add comments with the same syntax as described in Project 5.4. See Figure 1.7 for an example.

4. Save the file with a .kml extension and reopen it in Google Earth (see bold text in example KML code provided after this project).

Figure 5.3: Editing of KML code allows descriptions not to appear in the Places panel and omits default titles and directions from placemark balloons.

Projects 5.2–5.4   example code

```
<?xml version="1.0" encoding="UTF-8"?>
<kml xmlns="http://www.opengis.net/kml/2.2" xmlns:gx="http://www.google.com/kml/ext/2.2" xmlns:kml="http://www.opengis.net/kml/
2.2" xmlns:atom="http://www.w3.org/2005/Atom">
<Document>
        <name>Description.kml</name>
        <StyleMap id="msn_ylw-pushpin">
                <Pair>
                        <key>normal</key>
                        <styleUrl>#sn_ylw-pushpin</styleUrl>
                </Pair>
                <Pair>
                        <key>highlight</key>
                        <styleUrl>#sh_ylw-pushpin</styleUrl>
                </Pair>
        </StyleMap>
        <Style id="sh_ylw-pushpin">
                <IconStyle>
                        <scale>1.3</scale>
                        <Icon>
                                <href>http://maps.google.com/mapfiles/kml/pushpin/ylw-pushpin.png</href>
                        </Icon>
                        <hotSpot x="20" y="2" xunits="pixels" yunits="pixels"/>
                </IconStyle>
        </Style>
        <Style id="sn_ylw-pushpin">
                <IconStyle>
                        <scale>1.1</scale>
                        <Icon>
                                <href>http://maps.google.com/mapfiles/kml/pushpin/ylw-pushpin.png</href>
                        </Icon>
                        <hotSpot x="20" y="2" xunits="pixels" yunits="pixels"/>
                </IconStyle>
                <BalloonStyle>
                        <text>$[description]</text> <!-- Project 5.3-hides directions & the default title in placemark balloons -->
                </BalloonStyle>
        </Style>
        <Placemark>  <!-- Project 5.4-comments for Projects 5.2 & 5.3 -->
                <snippet maxLines="0"></snippet>  <!-- Project 5.2-hides description text in the Places panel -->
                <name>Project 5.2_5.4_Burr Trail</name>
                <description>The Burr Trail traverses the steep frontal limb of the Waterpocket Fold in Capitol Reef National Park.</
description>
                <LookAt>
                        <longitude>-111.0184541629349</longitude>
                        <latitude>37.85141955131326</latitude>
                        <altitude>0</altitude>
                        <heading>-94.71699129715726</heading>
                        <tilt>41.05223703428219</tilt>
                        <range>1653.602789638165</range>
                        <altitudeMode>relativeToGround</altitudeMode>
                        <gx:altitudeMode>relativeToSeaFloor</gx:altitudeMode>
                </LookAt>
                <styleUrl>#msn_ylw-pushpin</styleUrl>
                <Point>
                        <altitudeMode>clampToGround</altitudeMode>
                        <gx:altitudeMode>clampToSeaFloor</gx:altitudeMode>
                        <coordinates>-111.0226329570657,37.85112675782791,0</coordinates>
                </Point>
        </Placemark>
</Document>
</kml>
```

Project 5.5   programming time animations

1. In the **Places** panel, click RMB on the folder containing the features that you want to animate. Select **Save Place As...**, and a standard **Save** dialog appears.

2. Select **KML** from the **Format** drop-down menu.

3. Open the KML file with a plain-text editor like TextEdit (Mac) or NotePad (PC). Add the **TimeSpan** tags with **begin** and **end** tags after the **description** tag of each feature. If a feature is to be left on until present (i.e., activated at a certain time and then left on), it does not need an end tag.

4. Save the file with a .kml extension and reopen it in Google Earth. When the folder is checked, the time span control will appear in the upper left-hand corner so that you can view folder items based on their time stamps.

```
<Folder>
    ...
    </description>
    <TimeSpan>
       <begin>-Year-Month-Day</begin>
       <end>-Year-Month-Day</end>
    </TimeSpan>
    ...
</Folder>
```

5. In the KML example that follows, three placemarks were created and placed in a folder. The three placemarks highlight rock units in the Henry Mountains and Capitol Reef National Park areas that were formed in the Cenozoic, Mesozoic, and Paleozoic Eras, respectively. All of the KML code was automatically generated by Google Earth, except for the addition of the three TimeSpan entries (bold). For this example, we used the years to represent millions of years and a negative sign (-) to denote the past. Each placemark will appear and disappear as specified by its begin and end tags. A placemark first will appear during the period from 542 to 245 million years ago (~Paleozoic). Then the Paleozoic placemark will disappear, and a second placemark will appear during the period from 245 to 65 million years ago (~Mesozoic). Finally, the Mesozoic placemark will disappear, and a third placemark will appear during the period from 65 to 1 million years ago (~Cenozoic).

Project 5.5 example code

```xml
<?xml version="1.0" encoding="UTF-8"?>
<kml xmlns="http://www.opengis.net/kml/2.2" xmlns:gx="http://www.google.com/kml/ext/2.2" xmlns:kml="http://www.opengis.net/kml/2.2" xmlns:atom="http://
www.w3.org/2005/Atom">
<Document>
        <name>Project5.kml</name>
        <StyleMap id="msn_ylw-pushpin">
                <Pair>
                        <key>normal</key>
                        <styleUrl>#sn_ylw-pushpin0</styleUrl>
                </Pair>
                <Pair>
                        <key>highlight</key>
                        <styleUrl>#sh_ylw-pushpin</styleUrl>
                </Pair>
        </StyleMap>
        <Style id="sn_ylw-pushpin">
                <IconStyle>
                        <scale>1.1</scale>
                        <Icon>
                                <href>http://maps.google.com/mapfiles/kml/pushpin/ylw-pushpin.png</href>
                        </Icon>
                        <hotSpot x="20" y="2" xunits="pixels" yunits="pixels"/>
                </IconStyle>
        </Style>
        <Style id="sh_ylw-pushpin">
                <IconStyle>
                        <scale>1.3</scale>
                        <Icon>
                                <href>http://maps.google.com/mapfiles/kml/pushpin/ylw-pushpin.png</href>
                        </Icon>
                        <hotSpot x="20" y="2" xunits="pixels" yunits="pixels"/>
                </IconStyle>
        </Style>
        <Style id="sh_ylw-pushpin0">
                <IconStyle>
                        <scale>1.3</scale>
                        <Icon>
                                <href>http://maps.google.com/mapfiles/kml/pushpin/ylw-pushpin.png</href>
                        </Icon>
                        <hotSpot x="20" y="2" xunits="pixels" yunits="pixels"/>
                </IconStyle>
        </Style>
        <Style id="sh_ylw-pushpin1">
                <IconStyle>
                        <scale>1.3</scale>
                        <Icon>
                                <href>http://maps.google.com/mapfiles/kml/pushpin/ylw-pushpin.png</href>
                        </Icon>
                        <hotSpot x="20" y="2" xunits="pixels" yunits="pixels"/>
                </IconStyle>
        </Style>
        <StyleMap id="msn_ylw-pushpin0">
                <Pair>
                        <key>normal</key>
                        <styleUrl>#sn_ylw-pushpin</styleUrl>
                </Pair>
                <Pair>
                        <key>highlight</key>
                        <styleUrl>#sh_ylw-pushpin0</styleUrl>
                </Pair>
        </StyleMap>
        <Style id="sn_ylw-pushpin0">
                <IconStyle>
                        <scale>1.1</scale>
                        <Icon>
                                <href>http://maps.google.com/mapfiles/kml/pushpin/ylw-pushpin.png</href>
                        </Icon>
                        <hotSpot x="20" y="2" xunits="pixels" yunits="pixels"/>
                </IconStyle>
        </Style>
        <StyleMap id="msn_ylw-pushpin1">
                <Pair>
                        <key>normal</key>
                        <styleUrl>#sn_ylw-pushpin1</styleUrl>
                </Pair>
                <Pair>
                        <key>highlight</key>
                        <styleUrl>#sh_ylw-pushpin1</styleUrl>
                </Pair>
        </StyleMap>
```

```
<Style id="sn_ylw-pushpin1">
        <IconStyle>
                <scale>1.1</scale>
                <Icon>
                        <href>http://maps.google.com/mapfiles/kml/pushpin/ylw-pushpin.png</href>
                </Icon>
                <hotSpot x="20" y="2" xunits="pixels" yunits="pixels"/>
        </IconStyle>
</Style>
<Folder>
        <name>Project 5.5</name>
        <visibility>0</visibility>
        <Placemark>
                <name>Cenozoic</name>
                <visibility>0</visibility>
                <description>Tertiary intrusion that formed the Henry Mountains in Utah.</description>
                <TimeSpan>
                        <begin>-065-12-31</begin>
                        <end>-001-12-31</end> <!--don't need an end tag if it goes to present-->
                </TimeSpan>
                <LookAt>
                        <longitude>-110.9554208566362</longitude>
                        <latitude>37.9176967512585</latitude>
                        <altitude>0</altitude>
                        <heading>-27.44322278882175</heading>
                        <tilt>45.82052918099446</tilt>
                        <range>42374.69977423045</range>
                        <altitudeMode>relativeToGround</altitudeMode>
                        <gx:altitudeMode>relativeToSeaFloor</gx:altitudeMode>
                </LookAt>
                <styleUrl>#msn_ylw-pushpin</styleUrl>
                <Point>
                        <altitudeMode>clampToGround</altitudeMode>
                        <gx:altitudeMode>clampToSeaFloor</gx:altitudeMode>
                        <coordinates>-110.7988909377261,37.94668383465456,0</coordinates>
                </Point>
        </Placemark>
        <Placemark>
                <name>Mesozoic</name>
                <visibility>0</visibility>
                <description>Jurassic Navajo Sandstone along the steep frontal limb of the Waterpocket Fold.</description>
                <TimeSpan>
                        <begin>-245-12-31</begin>
                        <end>-065-12-31</end>
                </TimeSpan>
                <LookAt>
                        <longitude>-110.9554208566362</longitude>
                        <latitude>37.9176967512585</latitude>
                        <altitude>0</altitude>
                        <heading>-27.44322278882175</heading>
                        <tilt>45.82052918099446</tilt>
                        <range>42374.69977423045</range>
                        <altitudeMode>relativeToGround</altitudeMode>
                        <gx:altitudeMode>relativeToSeaFloor</gx:altitudeMode>
                </LookAt>
                <styleUrl>#msn_ylw-pushpin0</styleUrl>
                <Point>
                        <altitudeMode>clampToGround</altitudeMode>
                        <gx:altitudeMode>clampToSeaFloor</gx:altitudeMode>
                        <coordinates>-111.0237276810503,37.85352295017488,0</coordinates>
                </Point>
        </Placemark>
        <Placemark>
                <name>Paleozoic</name>
                <visibility>0</visibility>
                <description>Stream valleys cut down to expose the Paleozoic Kaibab Limestone. </description>
                <TimeSpan>
                        <begin>-542-12-31</begin>
                        <end>-245-12-31</end>
                </TimeSpan>
                <LookAt>
                        <longitude>-110.9554208566362</longitude>
                        <latitude>37.9176967512585</latitude>
                        <altitude>0</altitude>
                        <heading>-27.44322278882175</heading>
                        <tilt>45.82052918099446</tilt>
                        <range>42374.69977423045</range>
                        <altitudeMode>relativeToGround</altitudeMode>
                        <gx:altitudeMode>relativeToSeaFloor</gx:altitudeMode>
                </LookAt>
                <styleUrl>#msn_ylw-pushpin1</styleUrl>
                <Point>
                        <altitudeMode>clampToGround</altitudeMode>
                        <gx:altitudeMode>clampToSeaFloor</gx:altitudeMode>
                        <coordinates>-111.0471873204594,37.82153322522905,0</coordinates>
                </Point>
        </Placemark>
</Folder>
</Document>
</kml>
```

169

Project 5.6   finding useful Google Earth projects on the Web

- Click the **Earth Gallery** button in the upper right part of the **Layers** panel. Click the **Back to Google Earth** button in the upper left corner of the Google Earth viewer to return to Google Earth.

- David Rumsey Map Collection — http://www.davidrumsey.com/view/google-earth

- Google Earth Blog — http://www.gearthblog.com/

- Google Earth Community — https://productforums.google.com/forum/#!forum/gec

- Google Earth Cool Places — http://googleearthcoolplaces.com/

- Google Earth Design — http://googleearthdesign.blogspot.com/

- Google Earth Fractals — http://paulbourke.net/fractals/googleearth/

- Google Earth Hacks — http://www.gearthhacks.com/

- Google Earth Library — http://www.gelib.com

- Google Sightseeing — http://googlesightseeing.com/

- Maps Mania — http://googlemapsmania.blogspot.com/

- National Snow and Ice Data Center — https://nsidc.org/data/google_earth/

- Teaching with Google Earth — http://serc.carleton.edu/sp/library/google_earth/index.html

- USGS Earthquake data — http://earthquake.usgs.gov/learn/kml.php

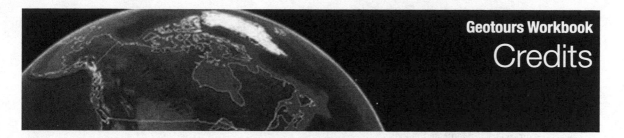

Cover

- Imagery (Front)-(top left) Google, CNES/Spot, DigitalGlobe, (bottom left) Google, CNES/Spot, DigitalGlobe, (bottom right) Google, U. S. Dept. of State Geographer, GeoBasis-DE/BKG, Data SIO, NOAA, U. S. Navy, NGA, GEBCO.
- Imagery (Back)-(top left) Google, Image Landsat, (middle left) Google, Data SIO, NOAA, U. S. Navy, NGA, GEBCO, ZENRIN, Image Landsat, (bottom left) Google, Image Landsat, (top right) Google, Image Landsat, Data SIO, NOAA, U. S. Navy, NGA, GEBCO, DigitalGlobe, TerraMetrics, (bottom right) Google, CNES/Astrium, DigitalGlobe.

Section 1: Getting Started
Part 1
Part 2

Section 2: Exploring Geology Using Geotours
A. Earth & Sky
- Eagle Nebula image-NASA/ESA/Hubble Heritage Team (STScI/AURA)/J. Hester, P. Scowen (Arizona State University).
- Pillars of Creation image-NASA/ESA/Hubble Heritage Team (STScI/AURA)/J. Hester, P. Scowen (Arizona State University).
- Crab Nebula image-NASA/CXC/SAO.
- Consolidated Impact Crater Database kmz file-KMZ file used with permission from Simon Levesque.
- Chesapeake Bay Impact overlay-Marshak, S., 2008, Earth: Portrait of a Planet, 3rd ed., W. W. Norton, 832 p.
- Impact Model image-Marshak, S., 2009, Essentials of Geology, 3rd ed., W. W. Norton, 518 p.
- Chicxulub Crater Gravity Map overlay-Reproduced with the permission of Natural Resources Canada, courtesy of the Geological Survey of Canada.

B. Plate Tectonics
- Iceland Geology overlay-Marshak, S., 2008, Earth: Portrait of a Planet, 3rd ed., W. W. Norton, 832 p.
- Seafloor Age Map-Seafloor age data downloaded from http://www.earthbyte.org/Resources/agegrid2008.html. Original data sources for age grid: Muller, R. D., Roest, W. R., Royer, J. Y., Gahagan, L. M., and Sclater, J. G.. 1997, Digital isochrons of the world's ocean floor, J. Geophys. Res., 102:3, 211-3, 214, & Müller, R. D., Sdrolias, M., Gaina, C., and Roest, W. R., 2008, Age spreading rates and spreading asymmetry of the world's ocean crust, Geochemistry, Geophysics, Geosystems, 9, Q04006, doi: 10.1029/2007GC001743.
- Tectonic Plates (plate names)-http://earthquake.usgs.gov/learn/plate-boundaries.kmz
- Tectonic Plates (plate boundaries)-http://earthquake.usgs.gov/regional/nca/virtualtour/kml/plateboundaries.kmz
- Earthquake overlay-Earthquake data downloaded from http://earthquake.usgs.gov/regional/neic/ and color-coded by depth (km) to create a Google Earth overlay by Dr. Tim Cope, DePauw University.
- Hawaiian Island ages-Problem adapted from a similar exercise developed by Dr. Tim Cope, DePauw University. Data adapted from a website by Ken Rubin (2005). Additional information available at: http://www.soest.hawaii.edu/GG/HCV/haw_formation.html. Additional information at: http://earthquake.usgs.gov/regional/neic/.
- Sag Pond image-Marshak, S., 2009, Essentials of Geology, 3rd ed., W. W. Norton, 518 p.
- Offset Stream image-Marshak, S., 2009, Essentials of Geology, 3rd ed., W. W. Norton, 518 p.
- Hot Spot Trail Map overlay-Marshak, S., 2008, Earth: Portrait of a Planet, 3rd ed., W. W. Norton, 832 p.
- Yellowstone Calderas overlay-Marshak, S., 2008, Earth: Portrait of a Planet, 3rd ed., W. W. Norton, 832 p.
- Yellowstone Hot-Spot Trail overlay-Marshak, S., 2008, Earth: Portrait of a Planet, 3rd ed., W. W. Norton, 832 p.
- Yellowstone Hot-Spot Trail ages-Smith, R. B., and Siegel, L. J., 2000, Windows into the Earth, The Geologic Story of Yellowstone and Grand Teton National Park, Oxford University Press, & Marshak, S., 2008, Earth: Portrait of a Planet, 3rd ed., W. W. Norton, 832 p.
- EMAG2: Earth Magnetic Anomaly Grid-Data downloaded from http://geomag.org/models/emag2.html. Original citation: Maus, S., Barckhausen, U., Berkenbosch, H., Bournas, N., Brozena, J., Childers, V., Dostaler, F., Fairhead, J. D., Finn, C., von Frese, R. R. B., Gaina, C., Golynsky, S., Kucks, R., Lühr, H., Milligan, P., Mogren, S., Müller, D., Olesen, O., Pilkington, M., Saltus, R., Schreckenberger, B., Thébault, E., and Caratori Tontini, F., 2009, EMAG2: A 2-arc-minute resolution Earth Magnetic Anomaly Grid compiled from satellite, airborne and marine magnetic measurements, Geochemistry Geophysics Geosystems, 10(8), Q08005, doi:10.1029/2009GC002471, ISSN: 1525-2027, http://geomag.org/info/Smaus/Doc/emag2.pdf.

C. Minerals

D. Igneous Rocks

- North America Batholiths Map overlay-Marshak, S., 2008, Earth: Portrait of a Planet, 3rd ed., W. W. Norton, 832 p.

E. Volcanoes

- Mt. St. Helens Volcanic Features overlay-Marshak, S., 2009, Essentials of Geology, 3rd ed., W. W. Norton, 518 p.
- 79 C.E. Eruption Time Sequence overlay-Adapted from Foss, P. W., Sigurdsson, H., and Robertson, S., 2007, Map 1 and Fig. 4.4, in J. J. Dobbins and P. W. Foss (eds.), The World of Pompeii, Routledge, 704 p. More information at: http://homepage.mac.com/pfoss/Pompeii/WorldOfPompeii/index.html and http://www.routledgearchaeology.com/books/The-World-of-Pompeii-isbn9780415475778.
- Hawaiian Island Landslides overlay-Marshak, S., 2008, Earth: Portrait of a Planet, 3rd ed., W. W. Norton, 832 p. (adapted from USGS/Barry W. Eakins).
- Mt. Rainier Hazards Map overlay-Marshak, S., 2008, Earth: Portrait of a Planet, 3rd ed., W. W. Norton, 832 p. (adapted from Fact Sheet by Scott, K. M., Wolfe, E. W., and Driedger, C. L., Hawaiian Volcano Observatory/U.S. Geological Survey).
- Mt. Rainier Features Map overlay-Overlay courtesy of the USGS (Topinka, USGSICVO, 1997; modified from Scott, et al., 1992). More information at: http://vulcan.wr.usgs.gov/Volcanoes/Rainier/Maps/map_place_names.html.
- Mt. Rainier Gigapan 360° panorama-http://volcanoes.usgs.gov/volcanoes/mount_rainier/multimedia_map_files.html.
- Crater Lake, OR Map overlay-Overlay courtesy of the USGS (Ramsey, D. W., Dartnell, P., Bacon, C. R., Robinson, J. E., and Gardner, J. V., 2003, Crater Lake Revealed, U.S. Geological Survey Geologic Investigations Series I-2790). More information at: http://geopubs.wr.usgs.gov/i-map/i2790/.
- Contour Map overlays-Adapted from Google Maps.
- Pyroclastic Flow video-Video courtesy of Miyagi, I., Kawanabe, Y., Takada, A., Sakaguchi, K., and Takarada, S., 2007, A Collection of Unzen Video Clips, GSJ Open-file Report no. 469, Geological Survey of Japan, AIST. More information at: http://www.gsj.jp/GDB/openfile/files/no0469/contents0469/index.html.

F. Sedimentary Rocks

- Grand Canyon Geologic Map overlay-Overlay courtesy of the USGS (Billingsley, G. H., 2000, Geologic Map of the Grand Canyon 30' x 60' Quadrangle, Coconino and Mohave Counties, Northwestern Arizona, U.S. Geological Survey, Geologic Investigations Series I-2688, Version 1.0). More information at: http://pubs.usgs.gov/imap/i-2688/.
- Great Unconformity video-Video courtesy of: Wilkerson, M.S. (2011), personal video.

G. Metamorphic Rocks

- Wind River Mts., WY Geologic Map overlay-Geologic map overlay generated using Google Earth application downloaded from http://www.gelib.com/geologic-map-united-states.htm. Original citation for entire geologic map: King, P. B., Beikman, H. M., and Edmonston, G. J. 1974, Geology of the Conterminous United States at 1:2,500,000 Scale, United States Geological Survey, http://pubs.usgs.gov/dds/dds11/index.html.
- New England Metamorphic Zones overlay-Marshak, S., 2008, Earth: Portrait of a Planet, 3rd ed., W. W. Norton, 832 p.
- Global Map of Shields & Mt Belts overlay-Marshak, S., 2008, Earth: Portrait of a Planet, 3rd ed., W. W. Norton, 832 p.

H. Earthquakes

- Regional Plate Tectonics overlay-Marshak, S., 2009, Essentials of Geology, 3rd ed., W. W. Norton, 518 p.
- New Madrid, MO Seismicity overlay-Marshak, S., 2008, Earth: Portrait of a Planet, 3rd ed., W. W. Norton, 832 p.
- Sag Pond image-Marshak, S., 2009, Essentials of Geology, 3rd ed., W. W. Norton, 518 p.
- Offset Stream image-Marshak, S., 2009, Essentials of Geology, 3rd ed., W. W. Norton, 518 p.
- 2004 Tsunami Travel-Time Animation image-Image courtesy of Kenji Satake at the Active Fault Research Center in Tsukuba, Japan, http://staff.aist.go.jp/kenji.satake/animation.gif.
- 2004 Epicenter image-Marshak, S., 2009, Essentials of Geology, 3rd ed., W. W. Norton, 518 p.
- North Anatolian Fault image-Marshak, S., 2009, Essentials of Geology, 3rd ed., W. W. Norton, 518 p. (adapted from Stein et al., 1996).
- Japan Seismogram (Problem 1)-Used with permission from the Virtual Courseware Project (ScienceCourseware.org) at California State University, Los Angeles.
- Japan Seismogram (Problem 1)-Used with permission from the Virtual Courseware Project (ScienceCourseware.org) at California State University, Los Angeles.
- Example Seismogram (Problem 1)-Used with permission from the Virtual Courseware Project (ScienceCourseware.org) at California State University, Los Angeles.
- Akita, Japan Seismogram (Problems 2,7)-Used with permission from the Virtual Courseware Project (ScienceCourseware.org) at California State University, Los Angeles.
- Example Seismogram (Problems 2,7)-Used with permission from the Virtual Courseware Project (ScienceCourseware.org) at California State University, Los Angeles.
- Busan, South Korea Seismogram (Problems 3,8)-Used with permission from the Virtual Courseware Project (ScienceCourseware.org) at California State University, Los Angeles.
- Example Seismogram (Problems 3,8)-Used with permission from the Virtual Courseware Project (ScienceCourseware.org) at California State University, Los Angeles.
- S-P Travel-Time Curve (Problem 4)-Used with permission from the Virtual Courseware Project (ScienceCourseware.org) at California State University, Los Angeles.

- Tokyo, Japan Seismogram (Problem 6)-Used with permission from the Virtual Courseware Project (ScienceCourseware.org) at California State University, Los Angeles.
- Example Seismogram (Problem 6)-Used with permission from the Virtual Courseware Project (ScienceCourseware.org) at California State University, Los Angeles.
- Richter Nomogram (Problem 9)-Used with permission from the Virtual Courseware Project (ScienceCourseware.org) at California State University, Los Angeles.
- NMSZ Structural Features-Data from Central Eastern United States Seismic Source Characterization for Nuclear Facilities Project, http:/www.ceus-ssc.com/index.htm.
- 1811-1812 Epicenters-Data from Central Eastern United States Seismic Source Characterization for Nuclear Facilities Project, http:/www.ceus-ssc.com/index.htm.
- NMSZ Liquefaction Features-Data from Central Eastern United States Seismic Source Characterization for Nuclear Facilities Project, http:/www.ceus-ssc.com/index.htm.
- North Anatolian Fault, Turkey-Fault traces and colored dates derived from: http://earthquake.usgs.gov/earthquakes/ eqarchives/year/1999/1999_08_17_ts.php and http://www.eqclearinghouse.org/2011-10-23-eastern-turkey/ 2011/10/31/geology-map-and-active-fault-map-for-turkey/.
- Tsunami Travel-Time Animation-National Centers for Environmental Information, http://www.ngdc.noaa.gov/hazard/ tsu.shtml.
- Tsunami Travel-Time Map-National Centers for Environmental Information, http://www.ngdc.noaa.gov/hazard/ tsu_travel_time_events.shtml.

I. Geologic Structures

- Geologic Map of the Bow Corridor, Canada overlay-Hamilton, W. N., Price, M. C., and Chao, D. K.,1998, Geology and mineral deposits of the Bow Corridor. Map 232, Alberta Geological Survey. http://www.ags.gov.ab.ca/ publications/ABSTRACTS/MAP_232.html. Used with permission from ERCB/AGS.
- Calgary to Castle Mtn Geologic Map, Canada overlay-Ollerenshaw, N. C.,1978, Geology, Calgary, West of Fifth Meridian, Alberta-British Columbia. Map 1457A, Geological Survey of Canada. http://apps1.gdr.nrcan.gc.ca/mirage/ show_image_e.php?client=mrsid2&id=1091003=gscmap-a_1457a_e_1978_mn01.sid. Reproduced with the permission of Natural Resources Canada 2009, courtesy of the Geological Survey of Canada (Map 1457A).
- Kananaskis Country Geologic Map, Canada overlay-McMechan, M. E.,1995, Rocky Mountain Foothills and Front Ranges in Kananaskis Country, West of Fifth Meridian, Alberta. Map 1865A, Geological Survey of Canada. http:// apps1.gdr.nrcan.gc.ca/mirage/show_image_e.php?client=mrsid2&id=2048973=gscmap-a_1865a_e_1995_mg01.sid. Reproduced with the permission of Natural Resources Canada 2009, courtesy of the Geological Survey of Canada (Map 1865A).
- Sheep Mountain Geologic Map overlay-Map reprinted by permission of the AAPG, whose permission is required for further use (AAPG © 2004): Banerjee, S., and Mitra, S., 2004, Remote surface mapping using orthophotos and geologic maps draped over digital elevation models: Application to the Sheep Mountain anticline, Wyoming: AAPG Bulletin, v. 88, no. 9, 1227-1237. Formation contacts modified from Hennier, J., 1984, Anticline, Bighorn Basin, Wyoming, Unpublished MS thesis, Texas A&M University, 118 p. Black strike & dip data are from Hennier (1984), whereas white strike & dip values are interpreted from DEM data by Banerjee & Mitra (2004).
- Split Mountain Anticline, UT Geologic Map overlay-Geologic map georeferenced in Google Earth by Dr. Tim Cope, DePauw University. Derived from Rowley, P. D., and Hansen, W. R., 1979, Geologic map of the Split Mountain quadrangle, Uintah County, Utah: U.S. Geological Survey, Geologic Quadrangle Map GQ-1515, scale 1:24000, and Rowley, P. D., Kinney, D. M., and Hansen, W. R., 1979, Geologic map of the Dinosaur Quarry quadrangle, Uintah County, Utah: U.S. Geological Survey, Geologic Quadrangle Map GQ-1513, scale 1:24000. Additional information at: http://pubs.er.usgs.gov/usgspubs/gq/gq1515.
- Geologic Map of Quail Creek SP, UT overlay-Original citation (with permission from the Utah Geological Survey): Biek, R. F., 2000, Geology of Quail Creek State Park, Utah, in Geology of Utah's Parks and Monuments, Sprinkel, D. A., Chidsey, T. C., and Anderson, P. B., eds., Utah Geological Association Publication 28. http:// www.utahgeology.org/uga_publist.htm.
- Arches NP Geologic Map-North overlay-Original citation (with permission from the Utah Geological Survey): Doelling, H. H., 2000, Geology of Arches National Park, Grand County, Utah, in Geology of Utah's Parks and Monuments, Sprinkel, D. A., Chidsey, T. C., and Anderson, P. B., eds., Utah Geological Association Publication 28. http://www.utahgeology.org/uga_publist.htm.
- Arches NP Geologic Map-South overlay-Original citation (with permission from the Utah Geological Survey): Doelling, H. H., 2000, Geology of Arches National Park, Grand County, Utah, in Geology of Utah's Parks and Monuments, Sprinkel, D. A., Chidsey, T. C., and Anderson, P. B., eds., Utah Geological Association Publication 28. http://www.utahgeology.org/uga_publist.htm.
- Geologic Map of Canyonlands NP overlay-Original citation (with permission from the Utah Geological Survey): Baars, D. L., 2000, Geology of Canyonlands National Park, Utah, in Geology of Utah's Parks and Monuments, Sprinkel, D. A., Chidsey, T. C., and Anderson, P. B., eds., Utah Geological Association Publication 28. http:// www.utahgeology.org/uga_publist.htm.
- Fin Garden video-Video courtesy of Wilkerson, M.S. (2011), personal video.

J. Geologic Time

- Grand Canyon Geologic Map overlay-Overlay courtesy of the USGS (Billingsley, G. H., 2000, Geologic Map of the Grand Canyon 30' x 60' Quadrangle, Coconino and Mohave Counties, Northwestern Arizona, U.S. Geological Survey, Geologic Investigations Series I-2688, Version 1.0). More information at: http://pubs.usgs.gov/imap/i-2688/.
- Geologic Map of Zion NP overlay-Original citation (with permission from the Utah Geological Survey): Biek, R. F., Willis, G. C., Hylland, M. D., and Doelling, H. H., 2000, Geology of Zion National Park, Utah, in Geology of Utah's Parks and Monuments, Sprinkel, D. A., Chidsey, T. C., and Anderson, P. B., eds., Utah Geological Association Publication 28. http://www.utahgeology.org/uga_publist.htm.
- Geologic Map of Cedar Breaks NM, UT overlay-Original citation (with permission from the Utah Geological Survey): Hatfield, S. C., and others, 2000, Geology of Cedar Breaks National Monument, Utah, in Geology of Utah's Parks and Monuments, Sprinkel, D. A., Chidsey, T. C., and Anderson, P. B., eds., Utah Geological Association Publication 28. http://www.utahgeology.org/uga_publist.htm.
- Geologic Map of Grand Staircase, UT overlay-Original citation (with permission from the Utah Geological Survey): Doelling, H. H., and others, 2000, Geology of Grand Staircase-Escalante National Monument, Utah, in Geology of Utah's Parks and Monuments, Sprinkel, D. A., Chidsey, T. C., and Anderson, P. B., eds., Utah Geological Association Publication 28. http://www.utahgeology.org/uga_publist.htm.
- Geologic Map of the Kaiparowits Plateau, UT overlay-Original citation (with permission from the Utah Geological Survey): Doelling, H. H., and others, 2000, Geology of Grand Staircase-Escalante National Monument, Utah, in Geology of Utah's Parks and Monuments, Sprinkel, D. A., Chidsey, T. C., and Anderson, P. B., eds., Utah Geological Association Publication 28. http://www.utahgeology.org/uga_publist.htm.
- Geologic Map of Circle Cliffs, UT overlay-Original citation (with permission from the Utah Geological Survey): Doelling, H. H., and others, 2000, Geology of Grand Staircase-Escalante National Monument, Utah, in Geology of Utah's Parks and Monuments, Sprinkel, D. A., Chidsey, T. C., and Anderson, P. B., eds., Utah Geological Association Publication 28. http://www.utahgeology.org/uga_publist.htm.
- Geologic Map of Quail Creek SP, UT overlay-Original citation (with permission from the Utah Geological Survey): Biek, R. F., 2000, Geology of Quail Creek State Park, Utah, in Geology of Utah's Parks and Monuments, Sprinkel, D. A., Chidsey, T. C., and Anderson, P. B., eds., Utah Geological Association Publication 28. http://www.utahgeology.org/uga_publist.htm.
- Geologic Map of Flaming Gorge, UT overlay-Original citation (with permission from the Utah Geological Survey): Sprinkel, D. A., 2000, Geology of Flaming Gorge National Recreation Area, UT-WY, in Geology of Utah's Parks and Monuments, Sprinkel, D. A., Chidsey, T. C., and Anderson, P. B., eds., Utah Geological Association Publication 28. http://www.utahgeology.org/uga_publist.htm.
- Geologic Map of Antelope Island SP, UT overlay-Original citation (with permission from the Utah Geological Survey): Willis, G. C., Yonkee, A., Doelling, H. H., and Jensen, M. E., 2000, Geology of Antelope Island State Park, Utah, in Geology of Utah's Parks and Monuments, Sprinkel, D. A., Chidsey, T. C., and Anderson, P. B., eds., Utah Geological Association Publication 28. http://www.utahgeology.org/uga_publist.htm.
- Geologic Map of Dead Horse Point SP, UT overlay-Original citation (with permission from the Utah Geological Survey): Doelling, H. H., Chidsey, Jr., T. C., and Benson, B. J., 2000, Geology of Dead Horse Point State Park, Grand and San Juan Counties, Utah, in Geology of Utah's Parks and Monuments, Sprinkel, D. A., Chidsey, T. C., and Anderson, P. B., eds., Utah Geological Association Publication 28. http://www.utahgeology.org/uga_publist.htm.
- Geologic Map of Canyonlands NP overlay-Original citation (with permission from the Utah Geological Survey): Baars, D. L., 2000, Geology of Canyonlands National Park, Utah, in Geology of Utah's Parks and Monuments, Sprinkel, D. A., Chidsey, T. C., and Anderson, P. B., eds., Utah Geological Association Publication 28. http://www.utahgeology.org/uga_publist.htm.
- Arches NP Geologic Map-North overlay-Original citation (with permission from the Utah Geological Survey): Doelling, H. H., 2000, Geology of Arches National Park, Grand County, Utah, in Geology of Utah's Parks and Monuments, Sprinkel, D. A., Chidsey, T. C., and Anderson, P. B., eds., Utah Geological Association Publication 28. http://www.utahgeology.org/uga_publist.htm.
- Arches NP Geologic Map-South overlay-Original citation (with permission from the Utah Geological Survey): Doelling, H. H., 2000, Geology of Arches National Park, Grand County, Utah, in Geology of Utah's Parks and Monuments, Sprinkel, D. A., Chidsey, T. C., and Anderson, P. B., eds., Utah Geological Association Publication 28. http://www.utahgeology.org/uga_publist.htm.
- Stratigraphy of the Grand Canyon image-Marshak, S., 2009, Essentials of Geology, 3rd ed., W. W. Norton, 518 p.

K. Earth History

- Appalachian Mts., TN Geologic Map overlay-Geologic map overlay generated using Google Earth application downloaded from http://www.gelib.com/geologic-map-united-states.htm. Original citation for entire geologic map: King, P. B., Beikman, H. M., and Edmonston, G. J. 1974, Geology of the Conterminous United States at 1:2,500,000 Scale, United States Geological Survey, http://pubs.usgs.gov/dds/dds11/index.html.
- Global Paleogeographic Model overlay-Images created by Dr. Ron Blakey (images used with permission and cannot be used by others for commercial purposes without Dr. Blakey's permission). The time-animated Google Earth overlays were created by Tony Pack (Google Earth materials used with permission and cannot be used by others for commercial purposes without Tony Pack's permission).

174

L. Energy & Mineral Resources

- Major Known Oil Reserves polygons-Adapted from Marshak, S., 2009, Essentials of Geology, 3rd ed., W. W. Norton, 518 p.

M. Mass Movements

- Hawaiian Island Landslides overlay-Marshak, S., 2008, Earth: Portrait of a Planet, 3rd ed., W. W. Norton, 832 p. (adapted from USGS/Barry W. Eakins).
- Mt. St. Helens Volcanic Features overlay-Marshak, S., 2009, Essentials of Geology, 3rd ed., W. W. Norton, 518 p.

N. Stream Landscapes

- Channeled Scablands overlay-Marshak, S., 2008, Earth: Portrait of a Planet, 3rd ed., W. W. Norton, 832 p.
- Mississippi River Delta overlay-Marshak, S., 2008, Earth: Portrait of a Planet, 3rd ed., W. W. Norton, 832 p.

O. Oceans & Coastlines

- U.S. East Coast Sea Level Changes overlay-Marshak, S., 2008, Earth: Portrait of a Planet, 3rd ed., W. W. Norton, 832 p. (adapted from Kraft, 1973).
- Largest Tidal Variations in the World, Bay of Fundy, Nova Scotia-National Weather Service, http://www.srh.noaa.gov/jetstream/ocean/fundy_max.html

P. Groundwater & Karst Landscapes

- Everglades, FL-Previous Groundwater Flow overlay-Marshak, S., 2008, Earth: Portrait of a Planet, 3rd ed., W. W. Norton, 832 p.
- Everglades, FL-Present Groundwater Flow overlay-Marshak, S., 2008, Earth: Portrait of a Planet, 3rd ed., W. W. Norton, 832 p.
- Winter Park, FL sinkhole-Photo courtesy of the USGS. More information at: http://sofia.usgs.gov/publications/ofr/01-180/opandaction.html.

Q. Desert Landscapes

- Syncline Development image-Marshak, S., 2008, Earth: Portrait of a Planet, 3rd ed., W. W. Norton, 832 p.

R. Glacial Landscapes

- Glacial Moraines-Long Island, NY & Cape Cod, MA overlay-Marshak, S., 2008, Earth: Portrait of a Planet, 3rd ed., W. W. Norton, 832 p. (adapted from Tarbuck & Lutgens, 1996).

S. Global Change

- Köppen-Geiger Climate Classification overlay-Overlay courtesy of Wilkerson, M.S., and Wilkerson, M.B. Original GIS shapefile downloaded from http://koeppen-geiger.vu-wien.ac.at/ from the Department of Natural Sciences, University of Veterinary Medicine Vienna. Original citation: Kottek, M., Grieser, J., Beck, C., Rudolf, B., and Rubel, F., 2006, World Map of the Kppen-Geiger climate classification updated. Meteorol. Z., 15, 259-263. DOI: 10.1127/0941-2948/2006/0130.
- World Sea Level Trends-National Oceanic and Atmospheric Administration, http://tidesandcurrents.noaa.gov/googleearth.shtml.
- Climate Central Sea Level Layers-Used with permission of Climate Central, http://sealevel.climatecentral.org/maps/google-earth-video-global-cities-at-risk-from-sea-level-rise.
- Sea Level Rise, New York City-Used with permission of Climate Central, http://www.climatecentral.org/wgts/2015-lamm-sliders/2015LammSliders-Global-embedcode.html.
- Sea Level Rise, Sydney-Used with permission of Climate Central, http://www.climatecentral.org/wgts/2015-lamm-sliders/2015LammSliders-Global-embedcode.html

Section 3: Developing Your Own Geotours

1. Placemarks

- Mount Saint Helens image (Fig. 1.10, Fig. 1.11)-USGS, http://volcanoes.usgs.gov/vsc/images/st_helens/Mount-st-Helens-home.jpg.
- Mount Saint Helens video-Courtesy of USGS YouTube channel (http://www.youtube.com/user/usgs), United States Geological Survey, http://youtu.be/sC9JnuDuBsU.

2. Paths & Polygons

3. Image & Photo Overlays

- Chicxulub Crater Topography image (Figs. 3.2-3.5)-NASA/JPL, http://photojournal.jpl.nasa.gov/catalog/PIA03379.
- Chicxulub Crater Gravity Map overlay (Figs. 3.4-3.5)-Reproduced with the permission of Natural Resources Canada, courtesy of the Geological Survey of Canada.
- Half Dome photo overlay-Courtesy of Wilkerson, M.S. (2011), personal photo.

4. Tours & Historical Imagery

- Tour of Folder Placemarks-Placemarks courtesy of Wilkerson, M.S. (2011). Derived from Powell, J.W., 1997, The Exploration of the Colorado River and Its Canyons, Penguin Books, 397 p.

5. Leftovers